NO NETTLES REQUIRED

'Gardens (all gardens) are good for wildlife, and
encouraging wildlife is entirely compatible with ordinary
gardening, costs next to nothing and is almost completely
effortless. You may have wished all these things were true,
but never allowed yourself to hope that they actually are.
'Well, they are, and here is the proof . . .'

Ken Thompson is a plant ecologist. He is Senior Lecturer in the Department of Animal and Plant Sciences at the University of Sheffield, where he was a key member of the first of two 'Biodiversity of Urban Gardens' (BUGS) projects investigating the significance of urban gardens as habitats for 'natural' biodiversity.

Ken Thompson has written over one hundred articles for scientific journals and writes a regular column on the science of gardening for *Organic Gardening* magazine. He is also the author of the critically acclaimed *An Ear to the Ground: Garden Science for Ordinary Mortals.*

NO NETTLES REQUIRED

The Truth about Wildlife Gardening

Ken Thompson

eden project books

TRANSWORLD PUBLISHERS
61–63 Uxbridge Road, London W5 5SA
a division of The Random House Group Ltd
www.booksattransworld.co.uk

First published in Great Britain in 2006 by Eden Project Books
a division of Transworld Publishers

This paperback edition published in 2007

A CIP catalogue record for this book
is available from the British Library.

ISBN 9781905811144

Address for Random House Group Ltd companies outside the UK
can be found at: www.randomhouse.co.uk
The Random House Group Ltd Reg. No. 954009

The Random House Group Ltd makes every effort to ensure that
the papers used in its books are made from trees that have been
legally sourced from well-managed and credibly certified forests.
Our paper procurement policy can be found at:
www.randomhouse.co.uk/paper.htm

Typeset in 11/13pt Granjon by
Falcon Oast Graphic Art Ltd.

Printed in Great Britain by
Cox & Wyman Ltd, Reading, Berkshire.

2 4 6 8 10 9 7 5 3

CONTENTS

PREFACE

'Every stroll around the garden leaves me with a query: who nibbled that leaf? Where have all the ladybirds gone? What is that bee doing on the bare earth? To me this is the joy of having a garden – being able to satisfy curiosity and eventually to understand some of the intricacies of the living world.'
Jennifer Owen, *Garden Life*, 1983

WE ARE ALL WILDLIFE GARDENERS NOW. LIKE IT OR NOT, YOU share your garden with hundreds, if not thousands, of different flies, beetles, spiders, bees, wasps and other creatures, not to mention a few mice and birds. Collectively these animals are very numerous, but you are probably unaware of the existence of nearly all of them. This huge reservoir of biodiversity presents gardeners with a clear choice. You can assume that these creatures are all out to get you (or at least your cabbages or roses) and work to reduce their numbers. It isn't easy, but you can try. Alternatively, you can embrace, encourage and enjoy the wildlife in your garden. In so doing, you will not only be helping to conserve Britain's native wildlife but also adding an extra dimension of sight, sound and activity to your garden. With just a little patience, anyone can learn to appreciate the living, breathing ecosystem right outside their back door. In fact I should warn you: wildlife watching can become addictive.

For the potential wildlife gardener, however, there's a snag. Although wildlife gardening has never been more popular, and advice about how to do it has never been more abundant, the quality of much of this advice is questionable. An internet search for 'wildlife' and 'gardening' yields over two and a half million hits. No doubt there are sites out there that will tell that the key to successful wildlife gardening is almost anything you care to name, from Buddhism to witchcraft. In fact a search for 'wildlife', 'gardening' and 'Buddhism' yields 58,500 websites. 'Wildlife', 'gardening' and 'witchcraft' results only in a rather disappointing 16,100 sites.

But even at the more respectable end of the market, advice on wildlife gardening tends to suffer from at least one of two fatal flaws. First, it simply recycles what everyone else believes. That doesn't mean it's wrong, although believing something just because everyone else does has never been a very useful strategy, and wildlife gardening, like everything else, has its fair share of urban myths. Second, a lot of advice comes from people who are trying to sell you something. That doesn't mean it's wrong either, but it certainly means it's likely to be far from complete. In short, wildlife gardening is in that unsatisfactory state where anecdote, commercial imperatives and the opinions of a few self-appointed 'experts' are far more prevalent than genuine evidence. All disciplines go through such a period. Only thirty years ago, medicine was in much the same state. For example, clot-busting drugs have only been widely recommended for heart attacks since 1986, even though their benefits would have been clear from a proper study of the evidence as early as 1975. In medicine, things have improved, but conservation is still struggling to escape its anecdotal past. New research has just revealed that the method used to monitor numbers of tigers in India for the last thirty years doesn't work. Closer to home, it's only just been noticed that the 1.6 billion euros spent on agri-environment schemes in The Netherlands have had no beneficial effects on wildlife at all. Wildlife gardening, as conservation's younger and arguably less scientific cousin, is if anything in an even worse state. Anyone can express an opinion and there's no way of knowing who is wrong and who – if anyone – is right.

To be fair, the dearth of reliable advice on wildlife gardening is no more than a reflection of a shortage of basic information. We know nearly as much about the Amazon rainforest and the deep ocean as we do about the inhabitants of the average suburban garden. By and large, ecologists have preferred to study more 'natural' habitats, and the intrepid few who have ventured into towns and cities have concentrated on the more accessible habitats, such as parks and derelict land.

One of the few searchlights into this gloom, and an inspiration for the work described in this book, is the heroic study by Jennifer Owen of her own garden in Leicester. Starting in 1972, long before wildlife gardening was fashionable, Owen painstakingly monitored the diversity, abundance and yearly fluctuations of many groups of animals in her garden. In 1991, she pulled together all the data into a classic book, *The Ecology of a Garden: The First Fifteen Years*, published by Cambridge University Press. This is compulsory reading for anyone seriously interested in garden wildlife, although it's only fair to point out that it's really only for connoisseurs of things with six (or more) legs.

Ground-breaking and remarkable as Owen's work is, however, it is only one garden, and perhaps not an entirely typical one. If we are ever to know which gardens are best for wildlife (and why), we need to look at lots of gardens. It was against this background that a few of us at Sheffield University set about finding the money to carry out Britain's first large-scale survey of garden biodiversity. Fortunately, the major government funder of environmental research, the Natural Environment Research Council, was running a special programme on urban ecology and was prepared (to their eternal credit) to stump up the cash. The fruits of the resulting research project, Biodiversity in Urban Gardens in Sheffield (BUGS for short), make up most of the first half of this book.

Being in possession of some real information on garden wildlife for a change, I am in the unusual position of being able to ignore most of what everyone else thinks is good for wildlife. Neither do I pay much attention to the numerous purveyors of wildlife goodies. Instead I adopt the unusual approach of looking at the evidence, which leads me to some surprising conclusions. Chiefly these are that wildlife gardening is easier, cheaper and more attractive than almost anyone would have you believe. In fact you'll still be doing a lot for wildlife if you put this book aside right now and carry on gardening as you always have, although for very little effort you can achieve a great deal more.

WHAT IS GARDEN WILDLIFE?

IN 2003 THE ROYAL HORTICULTURAL SOCIETY (RHS) commissioned MORI to explore gardeners' attitudes to wildlife. The results revealed an extraordinary level of interest in wildlife gardening: 70 per cent of respondents thought that people should consider wildlife when maintaining their garden, and a third of the gardeners watched wildlife in their gardens. On the other hand, there was clearly a problem when it came to translating this concern into action. Only half the gardeners believed that they were currently doing all they could to encourage wildlife, and 38 per cent said they would do more if they knew how. Between a quarter and a half of gardeners provided food, nest boxes or water, but only 16 per cent deliberately used plants to attract wildlife.

These findings largely back up an earlier survey by Scottish Natural Heritage (SNH), the government's wildlife agency in Scotland, which went into more detail about gardeners' attitudes to wildlife gardening itself. SNH asked gardeners how they would go about attracting wildlife into their gardens, a question that revealed an interesting consensus: wildlife gardening really only works in big gardens and isn't relevant to town gardens; wildlife gardens are hard work and look untidy; and plants for wildlife have to be native wild flowers rather than garden cultivars – in short, a major obstacle to encouraging wildlife is that it's unattractive, expensive and hard work. For those of you who want to get back out into the garden and not waste any more time reading, I can tell you now that there is scarcely a grain of truth in any of these beliefs. In fact the rest of this book will explain that gardens (*all* gardens) are good for wildlife, and that encouraging wildlife is entirely compatible with ordinary gardening, costs next to nothing and is almost completely effortless. You may have wished that all these things were true, but never allowed yourself to hope that they actually are. Well, they are, and later I'll show you the proof.

However, before we go any further, we really have to agree on a fundamental question, which is what is garden wildlife? And when I say 'agree' I mean that I'm going to tell you what I think, and you are either going to agree with me, or else read a different book. Recently the BBC has devoted many hours of live programming, on more than one occasion, to garden wildlife, almost all of it starring foxes, badgers, hedgehogs and a few birds. How representative is this of the wildlife that actually lives in the average garden? The answer is not at all. In fact this view of garden wildlife is as biased as a week of holiday programmes devoted exclusively to ballooning in the Andes and trekking to the South Pole. The reason is simply that both numbers and diversity of large, furry or feathered creatures in the typical British garden are extremely low. The average urban gardener is likely to encounter just five or six mammals, maybe seven if they can tell the difference between house mice and wood mice. You will see more birds than this, but not many. If you're lucky, and if you watch long and carefully enough, and if you include the birds that you see flying over your garden, you might eventually see around forty different bird species.

Compare that with Jennifer Owen's monumental study of the wildlife in her rather ordinary garden in Leicester. She concluded that you might eventually encounter more than 8,000 species of insect alone in her garden, plus several hundred other miscellaneous spiders, millipedes, woodlice and other invertebrates. Moreover, for those who care about biodiversity, the birds and mammals in your garden will be, almost without exception, common and widely distributed. So are most of the insects, but not all. Owen's garden contained species never before recorded in Britain, and a handful never recorded before anywhere, so why shouldn't your garden too? Of course, you won't know about these rare species, unless you plan to make your life's work a study of Gasteruptid wasps or Conopid flies. However, I assume that if you are bothering to read a book on garden wildlife, you really are interested in *all* garden wildlife, and not just the stuff you can see from the kitchen window. In any case, if your garden is good for wildlife, the evidence will be more than apparent to a casual observer in the form of the bigger animals further up the food chain, even if you remain happily oblivious of the previously undiscovered parasitic wasp hunting in your gooseberry bushes.

Think of your garden wildlife as a pyramid, with a few species at the top and many more further down. It would be a mistake, however, to assume that insects form the base of the pyramid, or even come anywhere near it. Below them come yet more, even smaller creatures. Consider the nematodes, or roundworms. They are big enough to be visible with the naked eye (just), but mostly pass unnoticed. Yet they have a claim to be some of the most diverse and significant animals (both economically and ecologically) on the planet, and one species or other lives almost everywhere. One is commonly known as the beer-mat nematode, which gives you a clue to how ubiquitous it is. Many are predators, but most are parasites, and it has even been suggested that every other animal species on earth has at least one personalized nematode parasite, which means that nematodes must be at least as diverse as everything else put together. Humans have plenty of nematode parasites, and around a quarter of the world's population is infected by one species or another. They're not headline species, because generally they don't kill you; they just make you ill. Others do enormous damage by attacking ornamental and crop plants, while others attack crop pests. Nematodes that attack slugs and vine weevils are now exploited commercially as biological pest controls. But even nematodes pale into insignificance when we move down in size to the level of the single cell; as many as 90 per cent of all species may be bacteria, and they really do live everywhere. Some grow best at temperatures above the boiling point of water, while others live a kilometre down in the earth's crust, very slowly munching their way through solid rock. But don't worry: to a much greater extent than with the insects and other creepy-crawlies, you need neither to be consciously aware of these creatures nor to do anything special to provide for them. Most of these creatures, on which all else depends, are already perfectly happy in your garden. Follow the simple advice in the following chapters and you should make them even happier.

Thinking about life on earth in all its variety helps to put hedgehogs and sparrows in perspective, but the myopic view of garden wildlife as mammals, birds and frogs has another unfortunate consequence: it makes most of us feel inadequate. If, like me, you have never had a family of foxes living under your garden shed (perhaps with me it is because I don't have a shed), you may feel like some kind of failure. It's liberating to know that foxes are just one of several thousand different kinds of animals that might live in your garden, with no greater claim to importance than any of the others. Foxes? Who needs foxes when you can have a whole family of *Bombus pascuorum* (a bumblebee) nesting in your garden? Less trouble than foxes, and more use too. It also seems to me that concentrating on foxes and birds is more than a little patronizing. There seems to be an unconscious assumption that gardeners, sated on a diet of David Attenborough, killer whales, gorillas and Komodo dragons, are incapable of appreciating anything that isn't big and (preferably) dangerous. How welcome it was to hear recently of a lone walker being attacked by a badger. At last, Britain seems to have some really attention-grabbing wildlife! Of course, it turned out to be a tame badger that just thought its dinner was overdue.

In reality, the attentive wildlife gardener will always find some fascinating activity in their garden, night or day, rain or shine, fox or no fox. Warfare as subtle and intense as any Second World War submarine drama is taking place all around you. In a delphinium flower, a perfectly camouflaged crab spider waits, motionless, for an unwary bee. In a nearby marigold, another crab spider has just got lucky and is draining the life from a hapless butterfly. In her nest of grass roots and moss, a bumblebee queen keeps a careful eye on her extended family, constantly alert in case any of them is planning a *coup d'état*. A solitary wasp hauls off a paralysed spider to its nest in an old tree stump. In the vegetable plot, a parasitic wasp seeks out and destroys cabbage white butterfly larvae, while in the pond, a dragonfly larva chews thoughtfully on a tadpole. As for the life-and-death struggle taking place in the compost heap, don't even think about it.

Nevertheless, there are gardeners out there who prefer large animals with fur or feathers, and who care little for the other 99.99 per cent of the wildlife in their gardens. Even they, however, would do well to ponder what my granny always used to say, which is that if you look after the pennies, the pounds will look after themselves (or cents and euros, just in case we have all had a miraculous change of heart by the time you read this). Most of the large animals in your garden are predators and eat the smaller animals, which means insects and other invertebrates. Frogs, newts, snakes, slow worms, hedgehogs, badgers, foxes and bats eat insects, worms and slugs, and if your garden is short of such victuals, its complement of large carnivores will be similarly impoverished. Many familiar garden birds are also predators, at least some of the time. Robins, blackbirds and thrushes will take fruit, but mostly eat invertebrates. Wrens, dunnocks and blackcaps eat little else besides insects. Even birds that will take nuts and seeds from a bird feeder as adults often need insects to feed their young. A pair of blue tits needs 10,000 caterpillars and other insects to raise a single brood of young.

If they cannot find these in your garden they will go elsewhere or, worse still, they will try to nest in your garden but some of their young will die. Quite a few of the more charismatic creepy-crawlies in your garden are also predators – spiders, dragonflies, lacewings, hoverflies (as larvae), ladybirds and many other beetles spring to mind.

Think of your garden as a supermarket, and try to think like the manager. To keep your customers (that is, birds, frogs and hedgehogs) happy, you need a constant supply of a wide range of high-quality products. Except unlike Tesco or Wal-Mart, this is a magic supermarket, where the shelves are miraculously restocked every morning with fresh provisions, without your having to lift a finger. To pursue the analogy, imagine for a moment a garden where the war with plant pests has finally been won, and caterpillars, slugs and aphids have been banished. This is the gardening equivalent of Tesco at closing time on Christmas Eve, with nothing left except some mouldy sprouts and a few tins of anchovies. You can no more expect any satisfied customers in such a war zone than you could expect the first floor of your house to survive the destruction of the ground floor. Of course, you can replace a fraction of the missing food chain by importing mealworms (beetle larvae) from a commercial supplier. For those whose aversion to worms forbids even this expedient, there are always dried mealworms at about £4 per 25 grams. But seriously, why bother? Your garden, with only your very modest assistance, is one giant larder, stuffed with every possible kind of food for the thousands of different kinds of predators that, given half a chance, would love to live there. All they need is to be allowed to get on with it.

One more thing. It's only natural to feel a little proprietorial about the wildlife in your garden. You think of the fox that trots across your garden as your fox, of the visitors to your bird table as your birds. In fact you share all these larger animals with the entire neighbourhood. There are only 125 million breeding birds in the UK, and your personal share of these is just two. Many of your larger garden visitors will travel several kilometres during their lives, seeking food, shelter and mates. Of course, you can make a difference to their lives, and the harder you try, the bigger the difference. But ultimately you are only a small part of the world of any of these creatures. Much the same is true of a few of the larger and more mobile invertebrates, such as bees, dragonflies and butterflies. But for the overwhelming majority of the smaller invertebrates, including the huge number that do not fly, your garden is their universe, where they are born, live and die. These are *your* wildlife, absolutely dependent on how *you* manage *your* garden. If you take it seriously, the responsibility is awesome, but so are the rewards.

So, let's sum up. First, garden wildlife is all about creepy-crawlies, which outnumber the more obvious things in your garden by about a million to one. Second, because most of this less-visible wildlife is born, lives and dies in your garden, it really is yours. Third, although much of this wildlife passes quite unnoticed, a lot of it can be observed even on a casual stroll through your garden. You can rely on bees, butterflies, hoverflies, spiders and ladybirds to provide interest and enjoyment even when the hedgehogs are in bed and the blue tits are in the next street. Finally, insects and other invertebrates are a barometer of the health of your garden, and if they are healthy and numerous, everything else will be too. So let's get on and see just how ridiculously easy it is to keep our garden wildlife happy.

INVESTIGATING
GARDEN WILDLIFE

FROM JENNIFER OWEN'S EPIC FIFTEEN-YEAR STUDY OF HER Leicester garden, and a few other relatively small studies, we know that gardens are potentially wildlife oases. In fact it's hard to understand why Owen's garden hasn't already been declared a Site of Special Scientific Interest. But there's a crucial snag, and it's that use of the word 'potentially'. Owen's garden isn't that extraordinary, but it also isn't quite average, so we really don't know if the other 16 million gardens in Britain are as good as hers, or worse, or even better. Inevitably, of course, there are probably millions of gardens in all three categories, but we don't know *why*. In short, we can't say with any great confidence what are the essential features of a wildlife-friendly garden. Not that there is any shortage of advice, but most of it is highly specific and targeted on particular kinds of wildlife, especially the three Bs: birds, bees and butterflies. Not only that, but most wildlife gardening prescriptions fall into the 'bleeding obvious' category: dig a pond, put up a bird table, grow nectar-rich flowers. Of course, following this advice will certainly help, but as we saw in the last chapter, most of it's aimed at the tip of an extremely large iceberg. The real state of the wildlife in your garden depends on things for which there are no prescriptions (for example spiders and wasps), which you actively dislike (aphids), or which you didn't even know existed (springtails and pseudoscorpions).

Our BUGS project set out to reveal for the first time what actually lives in a wide range of private gardens and, we hoped, provide some insight into why some gardens have more of these creatures than others. Because we lived and worked in Sheffield, that's where the study was conducted. The first thing we had to do was to get over the culture shock of working on private gardens, something none of us had ever done before. There's quite a tradition of urban ecology, much of it German, but none of it concerned with gardens. It's almost as though gardens are ecologically invisible. Similarly, gardens tend not to appear in local authority plans or policies concerned with the environment. Because their ownership is fragmented, and they lie outside any form of statutory control, politicians and scientists alike seem to have decided for a long time that the best policy is to ignore them (although this attitude is changing). There were times over the following three years when we were to have considerable sympathy with this view.

The second thing we had to do was think of an acronym for the project, since it is well known these days that no scientific progress can be made without one. I've known occasions when this has been the hardest part of the whole project, but this time the solution was simple: BUGS (Biodiversity in Urban Gardens in Sheffield).

I can foresee some trouble with the word bugs, so let's try to nip it in the bud here. My dictionary defines a bug as (among other things) 'any small insect', and in a garden context I think that would be the most widely understood meaning. To a biologist, however, a bug is a member of the Hemiptera, a very large order of insects with around 1,700 species in Britain. Bugs are characterized by piercing mouth parts, like a tiny hypodermic needle, which they use to suck the juices out of plants or other animals. Some plant-feeding bugs are all-too familiar pests (for example aphids), but some (for example the familiar shield bugs) are both harmless and rather handsome insects. Predatory bugs are much less obvious; perhaps the best known, at least to gardeners with ponds, are the pond skaters and water boatmen. Bugs are mostly harmless to humans, but water boatmen will bite if handled. Bedbugs are, I'm glad to say, less familiar than they used to be.

Here I will use bug in the strict biological sense.

The rest of this chapter tells you what we did in the BUGS project, and why. If the details don't interest you, you can quite easily skip them and move on to what we found, which is described in the following three chapters.

Our first practical task was to find some gardens to study. Despite being surrounded by thousands of them, this is not as easy as it sounds. There's always the temptation to take the easy route of going via some recognized environmental organization. Most surveys of garden wildlife adopt this approach. Recent surveys by the Mammal Society used gardens of members, while the Royal Society for the Protection of Birds (RSPB) looks at gardens of RSPB members. We could, for example, have contacted the local

Wildlife Trust and asked them to find us some volunteers. Unfortunately, gardens selected in this way would have been useless, because many would have been self-consciously 'wildlife' gardens, and therefore completely unrepresentative of gardens in general. It's a bit like asking the Countryside Alliance to nominate some people for a survey on attitudes to hunting with dogs. Nor would gardens belonging to members of gardening societies necessarily be any more representative. For example, the average RHS member's garden covers an area of 2,300 square metres, which is more than ten times the national average and much bigger than any garden in our survey. In the end, we adopted a policy of semi-public appeals and personal contacts, to arrive at 161 volunteer gardeners. Far too many, but we wanted to be able to select gardens to cover all the things we thought might be important. Eventually the 161 were winnowed down to sixty-one study gardens – fewer than we would have liked, but as many as we thought we could manage. We also decided at an early stage to restrict our attention to back gardens. Front gardens are often too exposed for some of the things we wanted to do, and anyway many terraced houses don't have them and an increasing number have been transformed into car parks.

It's only when you sit down and really think about it that you realize just how much gardens vary. Hardly surprising, really – gardens are very personal things and every one of us is different, so why shouldn't gardens show the same variety? If, in the end, we wanted to be in a position to say which gardens are best for wildlife, we had to cover at least the main ways in which gardens vary. First, size. Our smallest garden was 32 square metres, the largest 940 square metres. Or to put it another way, from about the size of my living room up to about the size of an Olympic swimming pool. There are larger gardens in Sheffield, and smaller ones, but a study over this range of sizes is relevant to most people's gardens. The average size of our study gardens was 151 square metres.

Second, type of house. Does it matter if your garden lies behind a terraced, semi-detached or detached house? In fact our sample of sixty-one gardens was just over half semi-detached, just over a quarter detached and the rest (16 per cent) terraced. A sample that exactly reflected the distribution of house types in the city would have had fewer detached houses and more terraced. Third, age. Garden age is a slippery concept, especially in these makeover-obsessed times, so the best we could do was try to cover a wide range of house ages. Our oldest house was 165 years old, while our newest, at just five years, was too new to be on our Sheffield street map.

Fourth, location within the city, and in particular distance from the edge. In fact, distance from the edge of the administrative boundary of Sheffield is not a very meaningful quantity, since on the east Sheffield merges with Rotherham, while on the west it butts up against the Peak District National Park. We therefore measured the distance of each garden from the nearest area of open countryside, usually grassland or woodland. One or two gardens were immediately adjacent to countryside, while our most urban garden was at least 4 kilometres from the nearest open country. Finally, something you might not consider if you're unfamiliar with Sheffield, altitude. Sheffield is an extremely hilly city, and in winter it's quite common to see 5 centimetres of snow on the roofs of cars from the outlying parts, when there has been no snow at all in the city centre. Our gardens covered almost the whole range of altitudes available in the city, from 40 metres to 250 metres above sea level.

Of course, gardens vary in many other ways than those already mentioned, and our gardens reflected this variety. Some had ponds and compost heaps, others had neither. Some were surrounded by walls, others by fences or hedges. Some were level, others sloping, sometimes steeply. Some were open and sunny, others densely shaded. Some had no lawn at all, while one consisted entirely of lawn. Forty-four garden owners put out food for birds, nineteen had nest boxes and twenty-five admitted to using slug pellets. Finally, our gardens varied in ways that are hard to quantify but obvious when you see them. For instance, some were so carefully manicured that we half expected to be asked to wipe our feet before walking on the lawn, while others showed little or no evidence of 'gardening' in the conventional sense at all. One garden was home to the largest German shepherd dog I've ever seen. Another hazard was garden owners with tea and biscuits to spare, which sometimes meant a visit that should have been over in ten minutes took two hours. All in all, however, it was a privilege to see a cross-section of 'ordinary' gardens, something normally denied to everyone except balloonists and burglars.

So, we had our gardens. Now we had to decide what to do with them. First, we made a complete description and scale map of each one, including a list of all plants; areas of lawns, flower beds, vegetable plots, paths and all other types of garden usage; height of the plants; numbers of trees; presence of garages, sheds, greenhouses, decking, ponds and so on. We also did our best to describe the surrounding area, including (for 1 hectare centred on the target garden) the number of houses, areas of buildings, garden, road and pavement, total area of 'green space' (from aerial photos) and (from the 1991 census) human population density.

Now the hard part. Time to stop talking about biodiversity – we actually had to catch ourselves some, and then identify what we had caught. Catching insects and other invertebrates isn't too difficult, but catching them in a standardized way across sixty-one gardens certainly is. In the end we adopted three main sampling techniques, two on the ground and one in the air. First, pitfall trapping, which is merely a domestic version of the pit of sharpened stakes much favoured by Tarzan movies. A disposable plastic coffee cup is sunk in the ground, its top level with the soil surface, and partly filled with 50 per cent alcohol; a wooden lid raised on four nails keeps out the rain. Pitfall traps are quite good at catching and killing insects and other small animals that walk on the ground. Without the alcohol, the victims would either escape or eat each other.

Now you may think that the arrangement of an alcohol-filled container placed carefully in a flower bed sounds strangely familiar, and you'd be right. You may even have used one yourself. We're talking about slug traps. In fact our pitfall traps proved astonishingly effective at catching slugs, which wasn't quite the intention. In some particularly slug-prone gardens (such as mine), the trap was full to the brim with slugs long before the fortnightly emptying time. Such traps were not very effective (beetles and spiders just skipped lightly across the pile of decomposing bodies) and didn't smell very nice either. It's things like this that persuade some biologists to take up biochemistry. You can fill pitfall traps with preservatives that don't attract slugs as potently as alcohol, but generally the possible replacements are more toxic, and we had to be careful in gardens that might be used by pets or small children. We sampled three pitfall traps in every garden throughout the summer and autumn of 2000.

Many insects live in the dead leaves and other debris (collectively litter) that accumulates on the soil surface, which is why you see blackbirds turning this stuff over so assiduously. So our other strategy for catching ground-dwellers was to collect samples of litter and extract the animals in them using Tullgren funnels. Tullgren funnels work by applying an intense (and hot) light to the top of some litter in a funnel. In trying to get away from the heat and light, animals in the litter make their way to the bottom of the funnel and fall into the usual pot of preservative. Tullgren funnels are an extremely effective way of persuading small animals hiding in leaf litter to come out with their hands up.

Our aerial sampling apparatus was the Malaise trap which, despite the name, is not a trap that makes you feel sick. It's named after Swedish entomologist René Malaise. Legend has it that Malaise returned after a hard day's collecting to find that his tent had caught more insects than he had, which inspired him to invent a trap based on a tent. Imagine a ridge tent with no side walls, a white roof and a central baffle and you get the general idea. Insects fly into the dark-coloured baffle, fly up towards the white roof and are funnelled towards the highest end of the ridge. Here there is a way out into a bottle of alcohol. Malaise traps are extremely effective at catching most kinds of flying insects.

Unfortunately Malaise traps are not exactly cheap, which made buying sixty-one out of the question. In any case, not all our survey gardens could easily have accommodated one. We had to content ourselves with twenty-four, one of which was promptly stolen. I still hope that somewhere, someone is still trying to figure out how they managed to steal a tent with no walls. Also, excellent as Malaise traps are, they have their limitations. For example, they are useless at catching butter-flies, for which there is really no substitute for lots of people armed with butterfly nets. We couldn't do this, so butterflies are missing from our study. Ditto moths. Moths can be caught very effectively in special night-time traps that use ultra-violet (UV) light as an attractant, but the traps are expensive and (worse still) need to be emptied every day before dawn.

Nevertheless, we felt we had done the best we could with insects of the crawling and flying variety. Unfortunately, this leaves out a large area in the middle: things that neither crawl on the ground nor fly much (if at all), but spend their time on plants and shrubs. For example, most spiders and many bugs and beetles. There are well-established ways of finding such things – you can simply search the vegetation, or you can place a tray beneath the plants and then whack them with a stick and catch what falls out – but the former is far too labour-intensive, and the latter would not have been popular with our volunteer gardeners. We tried a method that someone suggested to us, but that none of us had used before – the aerial pitfall trap. This is the usual alcohol-filled coffee cup, in a hole in the middle of a sheet of stiff card, which is supported among the vegetation on a couple of bamboo canes (you can tell all naturalists were *Blue Peter* fans in their youth). The idea is simple: insects walking on the card fall into the cup and are killed and preserved by the alcohol. Unfortunately, aerial pitfall traps are either a failed concept, or our inexpert technique simply wasn't up to scratch. For whatever reason, our traps caught a few bees and almost nothing else. Not recommended.

We did manage a good stab at one other group of insects – leaf miners. These are a mixed bag of moths, flies and beetles, all of which spend their larval period living inside leaves, producing an obvious 'mine'. Probably the most familiar example is the holly leaf miner, responsible for the brownish blisters you often see in holly leaves. Some garden ornamentals are quite prone to leaf miners, for example honeysuckle and aquilegias, and occasionally they can become a pest. Beetroot are sometimes badly affected. Crucially, leaf miners don't need catching, and most are reasonably easy to identify from the shape of the mine and the species of plant. If in doubt, you can always take the leaf home, wait for the adult to emerge and identify that.

Now, you might think that once we had filled a few thousand bottles with assorted pickled insects, the worst would be over. But you'd be wrong, because now we had to discover what we had caught. Ideally, exactly what we'd caught. In the same way that knowing how *many* birds visit your garden is a lot less interesting than knowing *which* birds visit your garden, we wanted to know more than just how many bees or beetles lived in our gardens. This was a problem, because although almost anyone can learn to identify spiders or wasps, given the will, the right books and equipment, abundant patience and all the time in the world, very few people actually do. One reason Jennifer Owen found several species new to science in her garden was the unlikely combination of someone being willing to spend years collecting a little-known group (ichneumonid wasps) and someone else having been prepared to spend years identifying them.

So, the next stage in the BUGS project hinged on finding experts who were able to do this job, and then persuading them to do it. Fortunately, we had foreseen this problem when applying for money in the first place, so we had some cash to offer as an inducement, but not a lot. If we had asked for enough money to pay our experts the commercial rate for their work, BUGS would never have been funded at all. Luckily, we had enough contacts between us to find someone to do most of the groups we hoped to cover. Note this is not the same thing as everything we had caught – we had far too little capital (financial or moral) to persuade anyone to identify the parasitic wasps and small flies hoovered up by the thousand by our Malaise traps. In the end, we found good homes for our collections of spiders and harvestmen, slugs and snails, beetles, bumblebees, solitary bees and wasps, sawflies, craneflies, hoverflies, bugs, centipedes, millipedes and woodlice. As well as having a huge taxonomic range, our captives included examples that illustrated most of the things that animals might be *doing* in our gardens: we had a good cross-section of animals that eat plants, animals that eat each other, animals that pollinate flowers and animals that either live in or eat (or both) leaf litter. Only parasitoids, far too diverse and difficult to identify, largely escaped our detailed attention.

We are all familiar with predators and parasites. Predators kill their prey (obviously) and a single predator usually kills several prey during its life. In contrast, parasites rarely kill their hosts, and a single host may support many parasites. Parasitoids have adopted a lifestyle that combines elements of both. They usually develop inside their hosts (as a parasite does), but always kill the host (as a predator does). Adult parasitoids search for their prey, usually the eggs, larvae or pupae of other insects, and then lay one or more eggs on, or more often in, the host. The young parasitoid develops inside the host, consuming it entirely apart from the skin, and then pupates and eventually emerges as a new adult. Most insect parasitoids are wasps, and the majority are small and rather inconspicuous. Indeed, the smallest of all insects are parasitoids. They are also diverse almost beyond belief, and not easy to identify, so I'm afraid we didn't, although we did count them.

Of course insects and other invertebrates also have ordinary parasites, just as mammals do. Do beetles have tapeworms? Yes, they do.

The result of all this hard work was a mountain of data. We had measured everything that could be measured in our sixty-one survey gardens. We knew as much as anyone is ever likely to know about the animals and plants that lived in those gardens. All we had to do was put these two things together and discover what (if anything) it all meant. The answer, or part of it anyway, is in the next chapter.

WHAT THE IDEAL WILDLIFE GARDEN DOESN'T NEED

THE BALD FACTS FROM THE BUGS SURVEY ARE STARTLING enough. We caught and counted about 40,000 individual invertebrates, including thousands of woodlice, slugs, bugs, beetles, bees, parasitoid wasps and various groups of flies. Many more were caught and not identified because it was simply too difficult and time-consuming to pigeonhole them even roughly. We were able to identify over 700 different kinds of animals all the way to the species level. Most of these were common, but not all. We found nationally rare beetles, bugs, snails and flies. A spider found in one garden was only previously recorded on mountain tops (an uninvited holiday souvenir?). A handful of species were the first or second record for Sheffield or Yorkshire, including the saxon wasp (*Dolichovespula saxonica*), which has colonized Britain in the last twenty years and only just reached Sheffield.

Mostly this confirmed what we already suspected – gardens are stuffed to overflowing with an enormous diversity of largely unseen wildlife. But these raw numbers don't tell us very much on their own. In wildlife terms, where do gardens stand in relation to, say, typical countryside? When, a few years ago, Dutch biologists looked carefully at the wildlife in fields of agricultural grassland, to see if taxpayers were getting value for money for the huge sums poured into agri-environment schemes designed to improve the wildlife value of farmland (it turns out they aren't, but that's a separate story), one group they examined in detail was bees. On average, the typical Dutch pasture is home to one or two species of bee. In almost all cases, this was the honeybee (there's only one kind of honeybee) and/or one of two very common species of bumblebee. How does that compare to gardens in Sheffield? Our worst gardens had five different bees, while our best had twenty. A typical garden had honeybees, six species of bumblebees and around three species of other (solitary) bees. Actually, this is not such startling news. It's already well known that gardens are good for bees. Bumblebee nests in gardens grow faster than those in farmland, even if the farmland is managed specifically to encourage wildlife. When Peter Spencer, chairman of the British Beekeepers Association, moved from rural Warwickshire to Birmingham, his yields of honey doubled.

Equally, we know that agricultural fields are rubbish for bees. Indeed, modern farmland might have been designed to make life difficult for bees. Bees are killed directly by insecticides, the weeds they depend on for food are killed by herbicides, and removal of hedgerows has eliminated potential nesting sites from large areas of farmland. In fact it's possible that bees, which are already extremely abundant in gardens, might do even better there if their numbers were not limited by the availability of nesting sites, but more about that later.

We know from other studies that what is true for bees is true for pollinating insects in general: that is, gardens are also good for butterflies, moths and hoverflies. For example, an intensive study revealed more than 600 species of moths and butterflies in the gardens of Buckingham Palace. Malham Tarn Nature Reserve in North Yorkshire, although much larger than Jennifer Owen's garden, seems to support only about the same number of species of hoverflies.

What about other kinds of invertebrates? The best comparative data are for animals that are easily sampled by pitfall trapping. Of these, the diversity of ground-dwelling beetles and spiders, for example, seems to be about the same for gardens and typical countryside. It's impossible to compare the diversity of animals that live in the plant canopy, such as web-spinning spiders, because we don't know much about those in gardens or in farmland. However, it would be very surprising indeed if gardens were not superior, since the diversity of these groups is strongly linked to the structural complexity of the vegetation, which is low in farmland and typically high in gardens.

So, gardens are certainly no worse for wildlife than farmland, and sometimes much better. Moreover, almost all kinds of farmland insects, from beetles to butterflies, have been declining steadily for decades and are continuing to do so. Modern intensive farming, with its reliance on high inputs of fertilizers, pesticides and herbicides, the loss of habitats such as hedgerows and ponds, and routine dosing of animals with drugs and antibiotics, have all conspired to reduce the abundance and diversity of invertebrates, with predictable knock-on effects on larger animals such as birds and bats.

But are *all* gardens good for wildlife, and what makes a really good wildlife garden? Crucially, what do our results tell us about the popular notions of wildlife gardening outlined in the previous chapter?

ONLY BIG GARDENS ARE GOOD FOR WILDLIFE

Careful readers will have noted that we did not scale our sampling to garden size. In other words, one garden, however large, contained just three pitfall traps and one Malaise trap. This is important because it bears directly on one of the things you might expect to influence garden wildlife, which is garden size. On one level, it's both obvious and trivial that, all things being equal, a garden twice the size of another will contain twice the number of insects, spiders, birds and so on. A much more interesting issue is whether large gardens are actually *better* for wildlife, above and beyond this simple effect of scale. In other words, if you sample a large garden and a small one in exactly the same way, with exactly the same number of traps, do you catch more animals (or more species, or both) in the larger garden? Now that is an interesting question, and the answer is no, no and thrice no, as Frankie Howerd might have said. Whether we looked at beetles or bugs, spiders or snails, and whether we looked at numbers of species or numbers of individuals, garden size stubbornly refused to appear as a significant factor in any of our analyses. Mature reflection reveals that any other outcome would have been extremely surprising, but first, what exactly does this mean? It does not mean that there is no more wildlife in a large garden than in a small one. Of course not, any more than there are no more kinds of breakfast cereal in a supermarket than in a corner shop. What it does mean is that if you apply the same, standard sampling effort (one Malaise trap, or three pitfall traps, or whatever) to gardens of different size, you catch the same in all of them. In other words, a small garden behaves just like part of a large garden.

Why is this not surprising? First of all, because if gardens weren't like this, they would be ecologically unusual. Whether looking at crustaceans or spiders, birds or mammals, ecologists generally find the same thing: if you select a small patch of habitat you find the same number of species and individuals per unit area as in a large patch. A recent study of bugs in Bracknell found that, using a standard sampling method, they couldn't distinguish between patches of grassland on traffic islands and the hectares of grass in the local park. In fact closely mown grass (in whatever quantity) was extremely bad for both numbers and diversity of grassland bugs. Long grass was much better, whether it came in small or large pieces.

Why is the world like this? Because insects, spiders and snails do not have the same attitude to property that you and I have. When we look at our gardens, it's difficult not to view the world as consisting of two utterly distinct parts – mine (or yours) and everyone else's. This proprietorial outlook is so ingrained that there are few things more irritating than a conflict at the boundary of neighbouring gardens. Nothing is more likely than a dispute over hedge height, overhanging trees or a broken fence to lead to thoughts of homicide, and occasionally even to murder. On the other hand, the animals that live in your garden are quite unaware of distinctions of ownership. They care not that this is your civilized plot, while over there is the wasteland owned by that oaf with the smelly bonfires and noisy children. A hedge, from their perspective, does not divide anything from anything; it's simply another piece of habitat. A fence is simply a place to hang a web, sunbathe or search for prey.

Before leaving the question of scale, it's worth noting that you can't always get rid of the effects of scale so easily. It's almost inevitable that larger animals (birds, say) will be observed at the garden scale. If you ask gardeners to record the birds that visit their garden, those with large gardens will see more birds than those with small gardens. Does this mean that large gardens are better for birds? That's another interesting question, and the answer is in a later chapter.

Forget any preconceptions you may have about garden size. Your efforts to attract wildlife will be rewarded in a tiny garden as much as they are in a large one.

TOWN GARDENS ARE USELESS FOR WILDLIFE

It's hard to shake off the feeling that the leafy suburbs are the best place for wildlife gardening. After all, where would you like to live if you were a butterfly? Of course, suburban gardens are often larger than inner-city gardens, but we've already seen that size isn't an issue. In fact the BUGS project revealed that distance from the edge of the city was no more important than size. None of the groups of invertebrates we looked at became more numerous or more diverse as we approached open countryside.

Like many slightly surprising results, this one becomes a lot less surprising once you think seriously about it. First, we already know that gardens are at least as good for wildlife as ordinary countryside. Given that this is true, why should there be more wildlife in places nearer the countryside than distant from it?

Second, the idea that garden wildlife gets more interesting as you near the edge of a city reflects a fundamentally wrong-headed view of the world. Garden wildlife is just that: wildlife that lives in gardens. A few, a very few, of the animals that visit your garden may be long-distance vagrants that strayed in from the nearest countryside. Maybe the odd bird or dragonfly. Mostly, however, the wildlife in your garden does not depend on continual immigration; it's there because it lives there, or at least in the immediate neighbourhood. If urban gardens contained only the last flotsam of a tide of wildlife that started out in the sticks, Buckingham Palace gardens wouldn't be home to 287 species of beetles alone. Nor would the Natural History Museum Wildlife Garden in London, created in 1995, already be home to over 300 species of moths. In the 1930s, Frank Lutz, Curator of Insects at the American Museum of Natural History, suggested that the museum raise his salary by ten dollars for every insect species over 500 that he could find in his average-sized New York City garden. To make the offer more enticing, he also volunteered to take a ten-dollar pay cut for every species that he fell short of 500. As it happened, the deal was not done, but Lutz probably wished it had been, since he eventually discovered 1,402 species of insect in his garden. Cities, even very big ones, are not wildlife deserts.

> *Do not give up on garden wildlife just because you live in the middle of a city. Most of the wildlife in your garden has more limited horizons than you have, and doesn't even know it lives in a city. And look on the bright side – in the middle of a city, you will not be troubled by the less welcome manifestations of rural wildlife, such as rabbits, badgers and deer.*

ARE NATIVE PLANTS BETTER FOR WILDLIFE?

It's worth spending some time on this one, because even those pundits whose wildlife-gardening advice is exemplary in every other respect still cannot resist recommending that you should grow native wild flowers. You've probably guessed by now that the BUGS project provided very little support for this view. Neither the numbers nor variety of any group of animals was much influenced by whether a garden had few or many native wild flowers. Recent German research, in which native and alien plants were grown experimentally and their wildlife compared, found exactly the same.

You will also not be surprised to learn that a moment's thought reveals that this outcome was not only likely, but inevitable. First of all, let's have a look at the idea of nativeness. Native plants are defined as those that have arrived in Britain entirely without human assistance. Sounds simple, doesn't it? Unfortunately it's not. A whole bunch of plants that were assumed to be natives when I was a student are now known, or strongly suspected, to be ancient introductions. In fact, modern botanical opinion on alien plants would gladden the heart of the most right-wing Home Secretary. Practically all our familiar cornfield weeds, including the common wild poppy, corn marigold and cornflower, are now regarded as aliens, even though many of them have been here since the Bronze Age. Never mind that large amounts of money and effort are now devoted to 'rescuing' declining rarities with evocative names such as pheasant-eye, shepherd's needle and Venus's looking-glass – they're all foreigners. Nor that some johnny-come-latelies such as the snowdrop, which was once thought to be native, may have arrived as recently as the sixteenth century. Such long-established aliens strike at the very heart of the idea that our native wildlife is somehow uniquely suited to life here, perfectly fitted to our climate and landscape.

That brings me to another peculiarly British problem with the whole alien/native dichotomy. There *are* places where it's OK to get worked up into a patriotic fervour about one's native flora. In New Zealand, Australia, Hawai'i and Madagascar, at least 80 per cent of the plants (and most of the animals) are endemic – that is, they live nowhere else on earth. The Japanese endemic flora is almost as large as the entire British native flora. Gardeners in such countries can legitimately take a pride in their native flora, most of which grows nowhere else. Despite the worldwide popularity of *Hebe*, *Phormium* and *Olearia*, New Zealand gardeners have spent decades trying to grow perfect roses and delphiniums but are now beginning to appreciate the potential of their native plants. Britain, however, is distinctly short of endemic plants, for the simple reason that almost our entire flora re-invaded from Europe only 10,000 years ago, which just isn't long enough for our plants to have become very different. All our 'endemics' are closely related to more widespread species, and most (if not all) would be better accommodated as sub-species. What's more, none is exactly common. Chances are you've never seen any, and wouldn't have noticed if you had. When did you last see Lundy cabbage, western ramping-fumitory or Welsh groundsel? You need patriotism of positively Churchillian proportions to get the British endemic list into double figures.

What this means is that virtually all British plants grow somewhere else and, with notable exceptions such as bluebell and heather, they're more common somewhere else too. So, across the water in France or Germany, you can find 'British' plants being eaten and pollinated by 'British' insects, all mixed up with plants and insects that didn't, for one reason or another, manage to get here. Another reason not to get too carried away by the idea of a uniquely well-adapted 'British' flora and fauna. Essentially there's no such thing.

In fact our views are strongly coloured by how well aliens appear to blend in and how much of a nuisance they make of themselves. Many farmers and some gardeners wage war on rabbits, while most of us have a soft spot for hares, but neither is a native animal. Generally speaking, the British are strongly in favour of the status quo, but do not particularly care how the status quo came about in the first place. Nobody cam-paigns for the eradication of fallow deer, which were unquestionably introduced, probably by the Normans. On the other hand, the difficulty of re-introducing the beaver, which was certainly a native animal until we exterminated it, has so far proved insuperable. As for bears, lynx and wolves, three other perfectly respectable natives, don't even think about it.

Plants of doubtful origin continue to turn up here. In 1989 the small-flowered tongue orchid (*Serapias parviflora*), previously known only in mainland Europe, was found growing in Cornwall. Orchids produce vast numbers of extremely tiny, light seeds, and it's entirely possible that they blew here across the Channel. If they did, the tongue orchid got here without human assistance, which means it qualifies as a native. As climate warming accelerates, more European plants might be encouraged to do the same. Then again, it's equally likely that seeds fell out of someone's trouser turn-ups or arrived stuck to the sole of someone's shoe. If that's how the tongue orchid got here, it's officially not a native and just another bloody weed. There is also a third possibility, which is that a rogue botanist planted the orchid deliberately, hoping to fool people into thinking it was a native plant. It wouldn't be the first time.

Arbitrary national boundaries further complicate the issue. Welsh poppy (*Meconopsis cambrica*) is familiar as a garden escapee throughout England, but it is truly native only in Wales. Scots pine is a native tree only in Scotland, and even there only in the Highlands. Every Scots pine in England and Wales is either planted or the descendant of a planted tree. It's bad enough if such boundaries stay put, but these days they often don't. The break-up of Yugoslavia and the former Soviet Union turned thousands of perfectly acceptable native plants into aliens, literally overnight. Not that being within the same political boundary is all that useful as a guarantee of native status; in the USA, prairie plants are sold as natives in New York and California, states that don't have any prairie and never did have.

The whole question of native and alien status gets fuzzier and fuzzier the closer you look, and if you look hard enough, it disappears completely in a puff of smoke. Nevertheless, let's suspend disbelief for a moment and imagine we live in a world where the slippery alien/native dichotomy can be nailed down satisfactorily. Native plants, we are told, are good for (native) wildlife, while aliens are not. The very paragon of nativeness in this respect, and the plant against which all others are measured and found wanting, is the English oak. A good choice – the very name conjures up a vision of solid respectability and good breeding. If there were a Burke's Peerage for plants, oak would be in it. Wildlife gardening books and websites tell us that oak supports hundreds of different kinds of native insect, many of which depend entirely on oak – they eat nothing else. The exact number depends on who you believe, and I suspect that the more remarkable totals are arrived at by adding together all the insects that eat *both* species of native oak, which is hardly fair. But let's not quibble. We can all agree that an impressive number of native insects live on oak trees.

What does oak tell us about native plants in general? The usual reasoning seems to run as follows: oak is native, oak supports lots of wildlife, therefore all natives support lots of wildlife. I'm sure philosophers have a name for this kind of logical fallacy, and one day I'll discover what it is. A more obvious example:

> All terriers are mammals
> All dogs are mammals
> Therefore all dogs are terriers.

Sometimes these logical errors are not so obvious, although it's usually easy to smell a rat even when you can't spot exactly where the rodent is hiding. For example:

> My mate Fred can't be a fool and a genius
> Fred is certainly no genius
> Therefore Fred is a fool.

See what I mean? In truth oak is a lot more than just native: it's also very common, very permanent and very, very big. In other words, for insects looking for a plant to eat, an oak tree is a very big target, so it's perhaps not surprising that so many have hit the bull's-eye. Sadly, not all native plants are either big or common, and even those that are don't always support much wildlife. Take bracken, which is possibly even more abundant than oak and equally native, yet supports hardly any insects at all.

However, both oak and bracken are red herrings, since most of us are unlikely to want to grow either in our gardens. Let's take a closer look at some natives you might actually want to grow. I've chosen, completely at random, five plants that satisfy three criteria. All five are pretty and certainly worth growing, they are undoubtedly British natives, and they are relatively uncommon as wild plants. In fact two of them, Jacob's ladder (*Polemonium caeruleum*) and maiden pink (*Dianthus deltoides*), are much commoner as garden escapees than they are as genuine natives. The other three are bird's-eye primrose (*Primula farinosa*), early purple orchid (*Orchis mascula*) and grass of Parnassus (*Parnassia palustris*). So, how many insect herbivores eat these five plants? It turns out that, as far as we know, one insect (a moth) eats grass of Parnassus and apparently nothing else, while the other four don't support a single specialist insect between them. Of course, entomologists have as much trouble finding uncommon plants as insects do, so there may be an element of under-recording here. However, none of our five native plants is difficult to identify, and all have showy flowers, so they can't have been completely overlooked by entomologists. One thing is plain: not all native plants support much (or indeed any) native wildlife.

Before leaving our five natives, one final point. Even if they support many more native insects than they appear to, there's another problem. To take one example, Jacob's ladder is confined (as a native plant) to parts of the Peak District and the Yorkshire Dales. If you garden in the south of England, what are the chances of the specialist herbivores of Jacob's ladder finding your plants? Anyway, wouldn't you secretly be happier if they didn't?

OK, you may be thinking at this point, maybe not all native plants support lots of wildlife, but some of them certainly do. If alien plants were uniformly poor for wildlife, it would still make sense, on average, to grow natives. Before we look properly at this, let's just think, for a moment, what proportion of garden wildlife depends closely on particular plants. We'll use the data from Owen's study because she has a more detailed list of garden wildlife than the BUGS study.

We are talking about invertebrates here, but it's worth mentioning in passing that the birds, mammals and amphibians in her garden cared little whether she grew native or alien plants. Most are generalist predators, and those that eat plants (for example, seed- or fruit-eating birds) neither know nor care about the origins of the seeds and fruit they eat. If they did, there would be no point in trying to feed them peanuts and sunflowers, would there? The berries of cotoneaster and pyracantha (both aliens) are avidly consumed by birds in my garden, but those of mezereon (a native) are usually not.

Owen identified at least 1,723 species of invertebrate animals in her garden. What are all these animals *doing*? More than half are predators or parasitoids, at least for part of their life cycle, and certainly have no great interest in the origin of the plants in the garden. Another large group are either generalist herbivores (for example, slugs and grass-hoppers), or eat dead plant or animal material (woodlice, millipedes, many flies), or subsist entirely on pollen and nectar (bees, and hoverflies as adults). This group also makes no distinction between native and alien plants. The rest, around a quarter of the total, are a mixed bag of herbivores – mainly beetles, moths, butterflies, sawflies and bugs. The activities of many, such as the gooseberry sawfly, large white butterfly and several species of aphids, are all too familiar to most gardeners. In their choice of food plants, they range from extreme specialists, confined to a single host plant (the small tortoiseshell butterfly on nettles), to those with a willingness to eat almost anything (the pretty angle shades moth, whose caterpillars will eat anything from lettuce to foxglove).

Much the largest group of garden herbivores is the moths, with 343 species in Owen's garden. Around 40 per cent are specialized to some extent, feeding on a single species, genus or family of plants, while the rest are less choosy. Moreover, the larvae of most moths (the familiar caterpillars) are fairly obvious, so Owen could tell what they actually ate in her garden. The results are illuminating. Of the 15 most popular food plants in the garden, each eaten by at least 5 different moth species, only 6 were native; 40 native plants (27 per cent of the total in the garden) were eaten by moth larvae, while 75 alien plants (35 per cent of the total) were eaten. Alien plants are far from a disaster for the caterpillars of native moths.

Why are so many native British insects ready to eat alien garden plants? One reason is that many alien plants are not all that foreign, and to see why we have to look at the history of the British flora. If we went back far enough, we would find all the earth's land stuck together in one big continent, Pangaea. About 130 million years ago, Pangaea broke in two. The southern half, Gondwanaland, later split up into the southern continents Australia, South America, Antarctica and India. However, the northern half, Laurasia, basically remained in one piece until quite recently. As a result there is a distinct northern temperate flora that's essentially the same everywhere. Botanists, who (like lawyers and doctors) aren't happy with an idea until they have a long name for it, call this the Arcto-Tertiary Flora. Thus, whether you're in Britain, New England or Japan, you'll find maples, oaks, birches, roses, asters and primulas – only the species differ. If you're a botanist and you know your plant families, floras from Seattle to Tokyo will all feel relatively familiar.

This is important because insects eat plants that taste right, which is a matter of chemistry, and chemistry runs in families. The majority of British garden plants come from North America, Asia, Japan and mainland Europe – regions that all share basically the same flora. In the average BUGS garden, 87 per cent of the alien plants came from families that also contain native species, while half even had native relatives in the same genus. This means that only the most choosy native insect herbivores will find nothing to eat in the average British garden. But you knew this already – surely you've noticed the perverse unwillingness of insect pests to eat (native) garden weeds, while tucking with relish into all your favourite garden plants? Note that things are very different in some other parts of the world. Australian insects, having had 50 million years to get used to eating eucalyptus, are unlikely to be impressed by roses and carnations.

But is it dangerous to generalize about the willingness of Lepidoptera to eat alien plants from the example of a single garden in Leicester? A recent study in California suggests not. Arthur Shapiro has spent a lifetime studying Californian butterflies and come to some startling conclusions about the butterflies of his home town of Davis. Davis is a medium-sized town, population about 60,000, and home to one campus of the University of California. Its butterfly fauna, intensely monitored since 1971, has a claim to be the best known in the USA. In Davis, 32 species of butterfly live and breed, all of them Californian natives except the ubiquitous cabbage white; 29 feed on alien plants at least some of the time, and 13 have no known native host plants in the town. The anise swallowtail, a spectacular and beautiful insect, depends almost entirely on fennel, a weed introduced from Europe. If fennel were eradicated, as is sometimes suggested, the citizens of Davis would lose their most iconic butterfly too. Shapiro has even shown that in one of the most remote places on earth, at the farthest tip of Patagonia, native butterflies are breeding on alien plants in towns.

To conclude, most wildlife is almost completely indifferent, for quite obvious reasons, to whether the plants in your garden are native or alien. Surprisingly, however, this applies even to many of the insects you might expect to depend more on native plants. But, and it's a big but, this does not mean that they don't care what you grow. It's no accident that gardens are exceptionally good habitats for pollinators. Bees appreciate the quantity and variety of flowers grown by gardeners, and in the BUGS study, the abundance and diversity of solitary bees were strongly linked to the total diversity of plants in particular gardens. The more different kinds of plants you grow, the more different kinds of pollinators will visit them. There are two reasons for this. One is that many pollinating insects are active for quite small parts of the year, and you need to have something in flower all the time if you are to attract the widest range of species. Another is that pollinators have preferences for particular flower shapes, sizes and colours. Generally speaking, flowers that are closest to the ancestral species are best for pollinators, while those that have been markedly modified by plant breeders are not so good.

One final point. Although I've concentrated here on the effects of alien plants on native herbivores, exactly the same applies higher up the food chain. All over North America, native birds, lizards and salamanders are perfectly happy to eat alien worms and insects.

> *Don't pay any attention to advice to grow native plants, because the wildlife in your garden probably won't. However, grow as many different flowering plants as you can, and make sure you cover the whole year, from* Mahonia *for the queen bumblebee that needs a snack on a warm day in February, to ivy for the butterflies that need one last fill-up before the winter.*

WILDLIFE-FRIENDLY PLANTS?

Nearly as widespread as a belief in the near-magical properties of native plants is the idea of wildlife-friendly plants – plants that provide good habitats or food sources for wildlife (personally, I always feel a bit sorry for the wildlife-*un*friendly plants). As usual there is no shortage of advice about which plants are friendly; indeed, you can buy whole books on the subject. In a narrow sense, wildlife-friendly plants certainly exist. A garden that consisted mainly of conifers and grasses wouldn't attract many pollinating insects, and if you want to attract fruit-eating birds, you need to grow shrubs with berries. Most of us can recognize plants with showy flowers or berries without any assistance, but it's sometimes useful to have your attention drawn to plants you might not have thought of. For instance, several native berrying shrubs or small trees, such as cornelian cherry (*Cornus mas*), sea buckthorn (*Hippophae rhamnoides*) and spurge laurel (*Daphne laureola*), are not grown nearly as often as they should be.

However, advocates of wildlife-friendly plants go well beyond these rather specific claims, suggesting that if you only grow enough of the right kinds of plants, your garden will become a paradise for everything from frogs to beetles. Before we can test these claims, we need an independent list of wildlife-friendly plants. Fortunately, the government's wildlife agencies in both England and Wales, Natural England and the Countryside Council for Wales (CCW), have each produced their own lists of wildlife-friendly garden plants. You can get both lists from their respective websites (www.ccw.gov.uk and www.naturalengland.org.uk).

The first thing that strikes you about these lists is how different they are. Both lists are about the same size, with around 160 or 170 plants, but they have only 45 plants in common. Obviously this wildlife-friendly business is not an exact science. To put it another way, they can't both be right. Or, to take a charitable view, if they are both right, they're both also very incomplete. The BUGS gardens turned out to be very variable in the numbers of plants they contained from either list. At one extreme, some of our gardeners grew hardly any of the plants on either list, while our most 'wildlife-friendly' gardens contained scores of plants from the combined list. Given this high level of variation between the BUGS gardens, any strong effect of wildlife-friendly plants on garden wildlife ought to show up clearly.

In fact, analysis of the data shows little if any relationship between any group of invertebrates and wildlife-friendly plants. But what exactly would we expect? Natural England is extremely vague about why the plants on their list are wildlife-friendly, but the CCW list is much more forthcoming. A few plants provide food or nesting sites for birds, but the overwhelming majority are there to provide nectar and pollen for pollinating insects. However, in the BUGS gardens we detected no significant effect of the CCW plants on the abundance of bumblebees, solitary bees, hoverflies or moths, or on the diversity of bees. This result has to come with a health warning: the pollinating insects expected to be favoured by wildlife-friendly plants are all flying insects that we sampled by Malaise trapping. Because we couldn't afford enough Malaise traps for every garden, our results were based on a relatively small number of gardens. Nevertheless, it seems likely that we would have detected any *strong* effect of wildlife-friendly plants on these insects. To put it another way, we would probably have detected any effect large enough to be noticed by the average wildlife gardener.

Which raises the obvious question: why does neither list of wildlife-friendly plants appear to be having the desired effect? In trying to answer this question, I think we need to look at the two lists separately because, frankly, the Natural England list is a bit of a dog's breakfast. First, it's almost completely opaque on the subject of why these plants are good for wildlife. I'm not saying its author didn't know why the plants are on the list, but such coyness doesn't fill me with confidence. Second, it relies very heavily on 'nativeness' as a criterion of 'friendliness' – more than two-thirds of the plants on the list are native wild flowers. Therefore the Natural England list doesn't work very well for much the same reasons that relying on native plants doesn't work very well. In fact, with its vagueness, spelling mistakes and reliance on native plants, the Natural England list looks very much as if it was put together late one Friday afternoon in a tower block in Peterborough. As of course it was (OK, I made up the part about Friday afternoon).

Perhaps I'm not being entirely fair to Natural England. Their job is to champion the conservation of England's native wildlife, so gardening is hardly their business. Having woken up quite recently to the wildlife potential of gardens, their views are still evolving. Another recent venture is a CD-ROM, *Gardening with Wildlife in Mind*, which lists 498 plants. Unlike the earlier list, this one is quite specific about the wildlife that is supposed to be attracted by every plant on the list. Unfortunately, it's also even more heavily biased in favour of native plants, which leads to two problems. First, some of the recommended plants are not only weeds but also downright ugly. You would have to be unusually dedicated to the welfare of the flame moth to bother growing knotgrass (*Polygonum aviculare*). Actually, the flame moth itself is no oil painting. The second problem is that many of these native plants are quite hard to find. Take the sand crocus (*Romulea columnae*), available from one nursery in Devon (and no mail order). Or the changing forget-me-not (*Myosotis discolor*). A pretty enough plant, as long as you have a magnifying glass handy, but quite unobtainable, even from specialist suppliers of wild flower seed. Equally unavailable is my favourite plant on the list, the pale butterwort (*Pinguicula lusitanica*), which is listed as attracting flies. So it should – it's insectivorous.

Am I still being unfair to Natural England? Yes I am. After

their earlier, less than successful forays into the arena of wildlife-friendly plants, it's clear that Natural England are now starting to get their act together. Their latest scheme is to persuade garden centres to label a short list of wildlife- (or at least pollinator-) friendly garden plants, although at the time of writing I'm afraid I can't tell you what these plants are. The BUGS project suggests that these plants are unlikely to make a big difference to the wildlife in your garden, but they will at least have the twin merits of being both good garden plants and widely available. I should also point out that most of Natural England's advice on other areas of wildlife gardening is exemplary. Their leaflets *Amphibians in Your Garden*, *Reptiles in Your Garden* and *Minibeasts in the Garden*, which can be downloaded free from their website, are excellent. Finally, I have to thank Natural England for providing us with a link from their website to the BUGS website (www.shef.ac.uk/uni/projects/bugs).

The Countryside Council for Wales seem to have got closer to a useful list of wildlife-friendly plants at the first attempt. Although their list contains a few native wild flowers, it's mainly composed of garden plants. Many are very good plants indeed, some of them specific cultivars such as *Aster × frikartii* 'Mönch', *Mahonia × media* 'Charity' and *Cistus* 'Silver Pink'. It also specifies single rather than double varieties only of *Aquilegia* and *Antirrhinum*. Every plant is there for a good reason: spindle for birds, fennel for hoverflies, red valerian for moths. In other words, it shows every sign of considerable thought, and looks as if it was put together by someone who wanted your garden not just to attract wildlife but to look good too. Even so, it doesn't seem to do what it says on the tin.

Why not? Well, part of the answer is easy. It doesn't work for, say, spiders or slugs because it was never intended to. Nearly all the plants on the list are there because they attract pollinators. The problem is that most of the plants favoured by gardeners attract pollinators. A typical cottage garden, designed with absolutely no regard for wildlife, would still be a paradise for pollinating insects. So it's not that the plants on the CCW list are not doing the job they were selected for, it's just that most of the plants in the average garden do the same job equally well.

Not only do lists of pollinator-friendly plants convey the (wrong) impression that plants not on the list are useless, but no list – however long – has any hope of being complete. In fact the rules for attracting pollinators are mainly common sense:

1. *Grow plants with lots of flowers. Don't overdo conifers, ferns, grasses and sedges.*

2. *Grow plants in large blocks, rather than as scattered individuals.*

3. *Make sure that there's something in flower right through the season, from hellebores, pulmonarias, aubretia, cherries and mahonias in spring, through to Michaelmas daisies,* Sedum spectabile *and ivy in the autumn. Try to put at least some of your spring flowers in a warm, sheltered spot.*

4. *Grow a good mixture of flower types: open and flat for beetles and hoverflies (for example, fennel, geraniums, saxifrages, spiraea), tubular and harder to access for large bees and long-tongued insects (foxgloves, campanulas, antirrhinums, broom and salvias), and night-scented for moths (nicotiana, honeysuckle).*

5. *Especially in a small garden, grow plants with a long flow-ering season, such as fuchsias, lavender, rosemary,* Potentilla fruticosa, Abelia *and* Hypericum *'Hidcote'.*

6. *Avoid plants with double or otherwise highly modified flowers.*

7. *Trust your own judgement. Make a note of plants that attract pollinators in your neighbours' and friends' gardens, or while out garden visiting, and grow them. If you already grow them, grow some more.*

Notice that these are mostly things the average gardener is probably doing anyway. In other words, you and the bees and but-terflies will get along fine if you garden simply to please yourself.

If, despite this advice, you're still looking for ideas for plants that look good and attract wildlife too, have a look at the CCW list. It's got plenty of good plants on it and you'll not go far wrong if you don't grow anything else. Or keep a look-out in garden cen-tres for the new Natural England wildlife-friendly bee symbol.

WHAT NEXT?

So we arrive at the end of a long list of things that turn out to have much less effect on wildlife than we have been led to expect. If that's all we found, you might by now be starting to wonder why we didn't just spend the BUGS money on a long holiday in the south of France. You might reasonably be ask-ing if the project found any positive results at all. After all, Sherlock Holmes thought it significant that the dog didn't bark in the night, but only because dogs often do bark in the night. The failure to find much effect of garden size and location, or of native or wildlife-friendly plants, becomes much more interesting if we found a positive effect of *something* on garden wildlife. Fortunately we did, and in the next chapter I'll tell you what it was.

LIES, DAMNED LIES AND COMPOST HEAPS

SUPPOSE I WERE TO TELL YOU THAT I COULD GUARANTEE TO toss a coin and get ten heads in a row. You'd be impressed, wouldn't you? You'd be less impressed if I told you I proposed to do this by tossing a coin repeatedly (a million times if necessary) until, by chance, I got ten heads in a row. In other words, you always have to judge a 'significant' result in the light of how many times you tried to get that result. In the long run, even highly improbable events become certainties, so saying that a roomful of monkeys with typewriters will eventually type all the works of Shakespeare is no more than stating the obvious. Most 'amazing' coincidences depend on ignoring how many opportunities the coincidence had to occur, but didn't.

So what is a 'significant' result? Biologists conventionally assume that something is significant if it would have happened by chance less than 5 per cent of the time. There's nothing magic about 5 per cent; it's just a convenient level of improbability. Of course, when a biologist gets a significant result, he doesn't actually know that he hasn't just got the freak positive result that you would expect by chance one time in twenty. Thinking something is statistically significant, when in fact it isn't, is a Type I error. You don't really need to know this – I'm just telling you so that you can show off the next time the conversation turns to statistics. Can anything be done about Type I errors? Yes, but at a cost. You can raise the bar to, say, 1 per cent, which means that you assume something is a real effect if it would have occurred by chance only 1 per cent of the time. No need to stop at 1 per cent either – you could say 0.1 per cent instead, which would make a Type I error very unlikely. But as you require your evidence to be less and less likely to be an accident, you run an increasing risk of missing a real effect, which statisticians call (you guessed) a Type II error.

An analogy is the standard of proof required in a civil and a criminal court. To be convicted of a criminal offence, a jury must think you are guilty beyond all reasonable doubt. This is a high standard of proof, because most of us would consider a Type I error (imprisoning an innocent man) to be worse than a Type II error (allowing the occasional criminal to go free). In a civil court, there is usually only someone's money at stake, rather than their liberty, so a lower standard of proof is reasonable; here the test is only the balance of probability. Thus it's perfectly possible to be found innocent in a criminal court but guilty of the same offence in a civil court. Ask O.J. Simpson.

Hmm ... very interesting, but what has all this to do with garden wildlife, you are asking yourself. Well, in our survey gardens we counted the number of species of many groups of invertebrates, and the number of individuals of many more groups. We also noted dozens of things about our gardens, such as size, age, type of house, position in the city, whether they had a pond, a compost heap, etc. So in the end there was a large pile of things to explain, and an equally large pile of things that might explain them. Putting the two together amounts to asking the same question over and over again. Whenever you do this, it's inevitable that you get some good correlations by chance. In fact, if your chosen level of probability is the customary 5 per cent, X will appear to be the cause of Y about one in twenty times, even if all your Xs and Ys are completely unrelated. Fortunately, many more than one in twenty of our results were significant; in other words it looks as if garden biodiversity really is related to measurable garden features. Nevertheless, it's inevitable that some of our results really are flukes, raising the difficult question: which ones? In separating the statistical wheat from the chaff, one can apply several criteria. First, repeatability. If the same pattern shows up many times, it's less likely to be an accident than something that happened only once. As Oscar Wilde noted, 'To lose one parent may be regarded as a misfortune; to lose both looks like carelessness.'

The BUGS survey contained only two really consistent patterns. The first, described in more detail below, was the positive effect of trees and shrubs on garden wildlife. This cropped up so often, in so many unrelated kinds of animals, that it has to be a real effect. To understand the second pattern, you have to know that Sheffield, perched on the edge of the Pennines, is a very hilly city indeed. The lowest part of the built-up area, in the Don Valley, is barely 40 metres above sea level, while the highest parts are 290 metres above sea level. Our survey gardens covered almost all this range. This altitudinal gradient is so obvious that it would be surprising if it had no effect on garden wildlife, and indeed it does. In fact, although it's not relevant here, we also found that altitude has a powerful (but largely unnoticed) effect on the plants growing in garden lawns.

Altitude does what you might expect. Many common invertebrates, such as beetles, spiders and wasps, were more abundant and diverse in the lower parts of the city, which are both drier and warmer. On the other hand, animals that prefer it cool and damp, such as slugs, snails, millipedes and craneflies, were more common in the higher, western parts of the city. Now you may say, with some justice, that this information is of little use to you unless you live in Sheffield. You may say it's not much use even if you do live in Sheffield, since you're hardly likely to move to another part of the city just to meet a better class of wildlife. On the other hand, there is a general lesson here for all gardeners. Generally warmth- and sun-loving animals like butterflies, grasshoppers, bush crickets, hummingbird hawkmoths and bee-flies are going to be easier to attract to your garden if you live in the south and east. The further north and west you live, the harder you will have to try to provide the warm, sheltered conditions that these insects like. Often, you may have to accept that you're unlikely to see them in your garden, just as you may be unable to grow some tender plants. Conversely, if you live somewhere cool and damp (as I do), you may just have to accept being overrun by slugs as part of nature's rich tapestry.

TREES

Compared to gardens with few or no trees, gardens with more trees or large shrubs had more beetles, slugs, snails, woodlice, millipedes, social wasps, sawflies, parasitoid wasps, craneflies and leaf-mining insects. This effect of trees stands out as by far the most dependable pattern from the BUGS survey. The diversity of wildlife that benefits from trees hints at the many ways in which trees benefit wildlife. First, trees literally add an extra dimension to gardens. Just as a five-bedroom detached house can accommodate a larger family than a two-bedroom terrace, trees increase the volume of living space available for wildlife. This means more plant-eating insects, and this effect cascades up the food chain. Great tits and blue tits may be familiar garden birds, but they raise fewer and smaller young in gardens than they do in woodland. The reason is simple – there's not enough to eat. Young great tits weigh just 1 gram, and need to increase that weight fifteen-fold in the two weeks after hatching. In a typical garden, adult tits often have great difficulty finding the huge numbers of caterpillars they need to keep their young alive. More trees make their job much easier. And if you're the sensitive type who can't bear to see your plants being consumed by aphids and caterpillars, the advantage of trees is that the damage all takes place well above eye level – out of sight and out of mind.

Trees also increase the variety of habitats in gardens. Bark is an insect habitat in its own right, but also provides space for lichens and mosses and thus for the animals that eat or inhabit these lower plants. If you're lucky enough to have an old tree, dead and decaying wood provides food and nesting sites for a huge variety of insects, from beetles to solitary bees. A hole in an old tree might attract hole-nesting birds such as tits, or even rarities such as bats. If your tree is really falling to pieces, you might even attract treecreepers. After all, nest boxes are only an attempt to make up for the shortage of natural tree holes in the modern landscape. Some trees, such as birches and alders, attract seed-eating birds such as redpolls and siskins.

Equally important is the indirect effect of trees on the garden environment. Gardens without trees, all things being equal, will tend to be relatively dry and sunny. For much garden wildlife this is fine, but many species, at least for part of their lives, prefer the damp, shady conditions provided by trees. The ground under a tree will also accumulate a layer of tree leaves in various stages of decay. These conditions will encourage a whole suite of animals that depend on damp, decaying plant material, such as earthworms, slugs, snails, woodlice, springtails and millipedes. They in turn will attract the animals that eat them, such as centipedes, spiders, ground and rove beetles and, ultimately, the birds, hedgehogs, frogs and shrews that eat them.

Trees are so important for garden wildlife that it's worth spending a few moments thinking about where they come from. I note that there is currently a campaign to get all UK children to plant a tree, which works out at 12 million trees. It's hard to imagine a more praiseworthy objective, and yet I can't help wondering where we got the curious idea that trees need to be planted. After all, not so very long ago Britain was almost completely covered by trees, and no one planted any of them. Seedlings of native trees and shrubs turn up in my garden all the time, mostly birch, willow, ash, yew, dog rose and holly, but also Scots pine, oak, hazel, hawthorn, elder and cherry. Aliens are less frequent, but include sycamore, sweet and horse chestnuts, cherry laurel, laburnum and assorted cotoneasters. The tree seedlings in your garden will be different, depending on local circumstances, but unless you live somewhere very remote from the nearest tree, a combination of wind, birds and squirrels will make sure that as soon as your back is turned, your garden will try to turn into a wood. To see what happens if we let nature take its course, consider the fate of Stocking Close Field, now better known (for reasons that will become obvious) as the Wilderness. This 4 hectare barley field next to Monks Wood National Nature Reserve, near Huntingdon in Cambridgeshire, was ploughed for the last time in 1961 and then abandoned. The only tree planting was undertaken by squirrels, jays and wood pigeons. By 1998 it was a 12 metre tall woodland, containing 891 oak trees, 582 ash trees and 1,481 other miscellaneous trees and shrubs, chiefly dogwood, hawthorn, blackthorn and dog rose. You may argue (rightly) that the Monks Wood wilderness had an existing wood next door, but consider another wilderness at Rothamsted Experimental Station in Hertfordshire. This former wheat field, surrounded by arable fields and pasture, was abandoned in 1882 and was quickly colonized by ash, sycamore, oak, hawthorn, dog rose, field maple and wild plum. It's now a wood of mature trees over 20 metres high. Plainly the landscape, including gardens, is full of young trees, and the only help they need to grow into big trees is to be left alone.

There's no great harm in planting trees, but it's worth remembering that a minimum of aftercare is essential. At least one in three of all planted trees dies before its fifth birthday, and some particularly careless planting schemes have achieved death rates of over 90 per cent. All but the smallest trees will require staking and watering during dry spells, while seedling trees require neither of these things. Planting trees can also encourage a careless attitude to nature. People are persuaded to plant trees, and to spend good money doing so, without even bothering to check whether there are young trees present already; young trees that appear naturally belong, by definition, to species that will grow well on that spot – their parents are proof of that. It's also a mistake to assume that anywhere can be improved by a spot of tree planting. Here in Sheffield, trees were mistakenly planted on a patch of old, species-rich grassland, because no one bothered to check what was growing there already. The result was the destruction of grassland worthy of preservation, and the replacement of it by a less than interesting wood.

'A garden without trees scarcely deserves to be called a garden.' Canon Henry Ellacombe, the nineteenth-century parson who said that, might not have had wildlife gardening in mind at the time, but he was completely right. As many have rightly pointed out, the ideal wildlife garden is an extension of the natural woodland-edge habitat, which means it needs trees.

Keep an eye open for self-sown tree seedlings in your garden, in friends' gardens, on waste ground and on country walks. Birch in particular produces prodigious quantities of seed and any bare ground near a mature birch will be carpeted with seedlings. Birch seedlings transplant easily and will give you a decent tree in ten years.

If you don't find what you want among your local tree seedlings, grow a tree from seed. It's very easy to collect seeds from many native trees such as field maple, birch, rowan and Scots pine. Our native rowan is one of the best small garden trees, but there are around seventy other rowan species in cultivation in Britain. Keep your eyes open: berries of many interesting species are easily picked up off the ground in gardens, parks and even the street. Remove the fleshy fruit, put the seed in a pot outdoors or in a cold frame, and prepare to be patient – germination may take two years. In fact, you'd be amazed what you can grow from seed. I have a Japanese maple, a mulberry, a tree peony and even a hand-kerchief tree (Davidia), all grown from seeds picked up off the ground in various gardens.

Even if, after all this exhortation, you still feel the urge actually to plant a tree, parting with £30 at your local garden centre should be your very last resort. Specialist nurseries will sell you young trees about a metre tall ('whips' in the trade jargon) for a pound or two. The Woodland Trust will sell you four different native trees for not much over a tenner, post and packing included (www.woodland-trust.org.uk). Such trees establish rapidly and soon catch up with larger and more expensive specimens. Moreover, they don't need staking.

OTHER USEFUL FINDINGS

Repeatability may be a good guide to *what* to believe, but there's also the important question of what is *worth* believing. I apologize for dragging us back to statistics, but there's a world of difference between a statistically significant relationship and a strong relationship between two things. For example, there is a well-established relationship between coronary heart disease and high blood cholesterol. But if you compare the range of cholesterol levels in people who have heart attacks and those who don't, they overlap almost completely. Only the tiny minority with very high blood cholesterol levels have a seriously raised risk of heart disease. For everyone else, other factors (smoking, obesity, diabetes, genetics, exercise) are much more important than blood cholesterol. In other words, although high blood cholesterol is highly significantly linked to heart disease (in a statistical sense), it explains very little of why some people have heart attacks and some don't. In much the same way, we obtained results for some invertebrate groups that, while statistically significant, left 90 per cent of the observed variation unexplained. This means that most of the differences between gardens were caused by something we didn't (or couldn't) measure (the equivalent of smoking or obesity in the cholesterol example), or were genuinely random. Either way it's not very helpful to the practical wildlife gardener. Another point to bear in mind is that things you can change are much more interesting than things you can't change. A few kinds of invertebrates seemed to be quite strongly influenced by garden surroundings. For example, social wasps (the big stingy sort) were more abundant in gardens surrounded by lots of green space (other gardens, playing fields, etc.). Bizarrely, parasitoid wasps were more abundant where there was less surrounding green space. Make what you will of either of these results, there is absolutely nothing you can do about either of them.

Finally, and perhaps most importantly, there is the question of plausibility. Some of our results made obvious biological sense, others did not. Solitary bees were more diverse in gardens with lots of different kinds of flowers. Gardeners who used slug pellets had fewer snails in their gardens (but not, sadly, fewer slugs) than those who did not. In each case, it's easy to see why we got this result. Cause and effect are obvious. On the other hand, numbers of hoverflies were positively related to both the number of houses and the number of people living in the immediate area. What is all that about? I haven't a clue.

So, what does all this add up to? If we run our results through a series of filters that remove weak effects, things that don't make sense and things we can't do anything about, we are left with a shortlist of things that are good, or bad, for at least some kinds of wildlife.

GOOD FOR WILDLIFE:
A wide variety of different plants and many different kinds of vegetation, for example lawn, flower beds, shrubs, trees, etc.
- *Ponds*
- *Compost heaps*
- *Hedges*
- *Walls* continued ▶

BAD FOR WILDLIFE:
- *Lots of hard surfaces, for example paving, decking*
- *Tidiness*
- *Slug pellets*

A glance at these lists reveals that this is not rocket science. In fact it's hardly First World War biplane science. Dull gardens are poor wildlife gardens. Gardens without ponds or compost heaps are poor wildlife gardens. Open, windswept gardens without walls or hedges (internal or external) are poor wildlife gardens. Gardens where plants take second place to car parks, patios and decking are poor wildlife gardens. None of this is surprising: gardens that provide a diversity of different kinds of food and places to live, and plenty of both, are good for wildlife. Perhaps only tidiness calls for particular comment. We were keen to get a feel for our garden owners' style of gardening, so we asked them about their attitude to weeding, pruning, watering, dead-heading and clearing up fallen leaves. In each case we asked them to rate themselves from 'zealous' at one extreme to 'laid back' at the other; then we added the scores together to get an index of intensity of garden management ('tidiness' for short). Finally, for good measure, gardeners who admitted using insecticides or herbicides got extra tidiness marks. Not surprisingly, a high score was bad for wildlife, but to see exactly why, you need to know something about how most ecosystems work.

If you were ever told anything about ecology at school, it

probably involved a 'food chain', in which grass was eaten by caterpillars, caterpillars were eaten by blue tits and the blue tits were eaten by a solitary sparrowhawk. In a more exotic version, the grass was eaten by wildebeest, which later provided lunch for lions. As a caricature of how grasslands work, neither of these descriptions is far from the truth. However, most of the world, and especially anywhere where trees grow, doesn't work like this. The fact is that most living plant material is not eaten by herbivores. Most stems and leaves die without being eaten, and the dead plant material then provides food for a huge army of decomposers. Earthworms, most slugs and snails, millipedes, woodlice and springtails are the most familiar decomposers, but many – for instance bacteria and fungi – are not animals at all. In a typical woodland, over 90 per cent of the energy captured by the trees goes via this decomposer route. Not surprisingly, an equally large army of spiders, centipedes, beetles and mites (plus more exotic predators, like pseudoscorpions) feasts on the decomposers, while the whole lot provides food for birds, frogs and hedgehogs.

It's therefore easy to see why every time you clear up fallen leaves or tidy up the herbaceous border, you are sabotaging one of the most important engines of diversity in your garden. Jennifer Owen realized that tidiness was one of the great enemies of garden wildlife, and described her strategy: 'Pruning and clearing is kept to a minimum, and delayed until winter whenever possible. Good ground cover is maintained, both by cultivated plants and, failing that, by weeds.' Oliver Rackham, the most astute modern chronicler of the changing English landscape, went even further and attributed much recent damage to the English countryside to excessive tidiness. This is Rackham on the modern fashion for indiscriminate trimming of hedges: 'Old overgrown hedges, full of blossom in spring and set with hollow ivy-tods and other reminders of antiquity, are part of the romance of the English landscape. We remember them from our childhood and see them in the pictures of Constable and Arthur Rackham and the verses of John Clare; we see them still in places that have escaped the Vandal hand of tidiness.' Tidy landscapes are lifeless landscapes, both aesthetically and biologically.

Something else that emerges from the list on page 72, although not as clearly as we would have liked, is the question of poisons. When the BUGS project began, we realized right away that it would be extremely useful to know if gardeners used pesticides. The second thing we realized was that there was little chance of getting reliable, accurate information about this. With the best will in the world, few gardeners would be able to tell us exactly which chemicals they used, how often and in what quantities. So finally we asked simply 'Do you use insecticides?' and 'Do you use herbicides?' Despite getting yes or no to these questions from most of our garden owners, we soon realized that this information wasn't precise enough to be worth analysing on its own. The only specific chemical we asked about was slug pellets. We felt this information was good enough to include in the analysis, and sure enough, it looks as if slug pellets actually do reduce numbers of snails, even if not slugs. However, apart from that, the BUGS project itself has little to say about the effect of pesticides. Nevertheless, it seems self-evident that attempting to poison the animals in your garden is not the best way to encourage them. Spraying chemicals to kill aphids and caterpillars will also kill many innocent bystanders, and even if it doesn't, it will persuade the myriad animals that eat them to go elsewhere. In any case, if you are prepared to read a book on wildlife gardening (as you are), you are clearly not a big user of the products of Monsanto, Bayer CropScience and Du Pont. So we'll agree on that and move on.

COMPOST HEAPS

Compost heaps and ponds both call for special comment. You don't need me to tell you that every garden should have a compost heap, if only because composting garden and household vegetable waste is the easiest recycling you'll ever do. And if you think your small heap doesn't make much difference, consider how much waste the nation's compost heaps deal with. The average Briton consumes about 2.2 kilograms (fresh weight) of fresh fruit and vegetables per week. This is the amount that actually enters the house, and it seems reasonable to assume that about a quarter is waste, in the form of potato peelings, apple cores, banana skins and so on. Thus a household of two would accumulate over 1 kilogram of waste every week from kitchen sources alone, while garden waste (such as weeds and lawn clippings) is probably at least the same again. Therefore if we assume that at least 2 kilograms of waste material is added to the average compost heap every week, this accounts in Sheffield alone for the processing of 5,278 metric tonnes of material every year, a tonnage that the city refuse services would otherwise have to handle and much of which would be disposed of by incineration or by landfill. This is good news, but we could do better. In Sheffield, which there is no reason to believe is unusual, only 29 per cent of gardens have compost heaps. That may be an impressive 50,750 compost heaps, but if every private garden had a compost heap, at least 18,200 metric tonnes of material would be recycled every year.

But what about compost heaps and wildlife? Unfortunately, the BUGS project had neither the time nor the money to look at compost heaps in detail, but other research (much of it from Scandinavia) reveals that compost heaps are hugely important wildlife habitats in their own right. This is because warm, decaying plant material is a habitat that would otherwise be absent from gardens. Reptiles in particular like these conditions, and a survey in Bristol found that gardens with compost heaps were twice as likely to have resident slow worms as those without compost heaps. Given the appetite of slow worms for slugs, this has to be another good reason for starting a compost heap. If you're really lucky, grass snakes might breed in your compost heap. There aren't any reptiles in my Sheffield garden, but wood mice breed in my compost heap. And please don't forget that both slow worms and grass snakes are completely harmless to humans.

In the BUGS survey, compost heaps had a small but detectable effect on garden wildlife, chiefly noticeable as a greater diversity of beetles in gardens with compost heaps. This is hardly surprising since, although all kinds of creatures live in compost heaps, beetles are compost animals par excellence. There have been relatively few studies of compost wildlife, but we know that British compost heaps are home to at least 300 species of beetle alone. Kings of the compost heap are the staphylinids, or rove beetles. All are predators and the devil's coach-horse, one of Britain's largest beetles, is the *T. rex* of the compost heap. Since compost heaps are a relatively recent, man-made habitat, you may wonder what all these creatures were doing before there were compost heaps to live in. In fact, most present-day compost inhabitants started out in a variety of more natural habitats, including decaying fungi, carrion, dung, dead wood and nests of birds and mammals. In other words, compost heaps go a long way towards replacing a whole catalogue of wildlife habitats that would otherwise be rare or absent from gardens, and are increasingly rare in the wider countryside. Another interesting feature of compost heaps is that by providing a warm, insulated habitat, they allow many species to spread further north than they would otherwise do. Many native species live further north in compost than in any other habitat, and alien species from warmer climates frequently survive best in compost heaps.

Home-made compost is one of the best soil improvers you'll ever have, and certainly the cheapest. Compost heaps are also a uniquely valuable wildlife habitat in their own right. No gardener with any interest in wildlife should be without one, however lacking you may be in the raw materials. Even if your garden is almost entirely covered by decking and you eat no fresh vegetables at all, empty pizza boxes make an excellent compost heap ingredient.

PONDS

Almost half the BUGS survey gardens had ponds – in fact some had more than one, so we were able to take a detailed look at thirty-seven ponds in all. We visited them during August and September 2000, collected samples to check what was living in them, and noted all the things we thought might be important – depth, area, clarity of the water, pH, plants, whether we saw (or garden owners reported seeing) fish, frogs or newts and so on. The first thing to report is that ponds, like gardens, are teeming with life. We found lots of different insects (beetles, bugs, dragonfly larvae, backswimmers, pond skaters, caddis fly and mayfly larvae), plus water snails, fresh-water mussels, leeches, flatworms and water slaters, although no single pond had very many of these creatures. Many ponds had frogs and nearly as many had newts. In other words, garden ponds are diverse ecosystems, consisting of many of the different kinds of animals you would expect to find in farm ponds.

But they weren't all the same, so why are some ponds more diverse than others? Two things stood out: plants and fish. Ponds with lots of submerged plants had more kinds of animals than those with little or no vegetation. There are probably many reasons for this. Plants provide hiding places from predators and places to lay eggs, and help to keep the water oxygenated. The effect of fish was more interesting. The numbers of different kinds of animals did not differ in ponds with fish or without, but fishy ponds generally had lower numbers of individuals. Some animals didn't like fish at all – we found few water fleas (*Daphnia*) in any pond that also contained fish. Given that water fleas help to keep the water clear and are food for a host of other pond animals, this is clearly a problem. A more subtle effect was that the animal communities in ponds with fish were all rather similar, while fish-free ponds were much more variable. In a nutshell, ponds with fish are dull (unless you like fish, of course), while ponds without fish tend to be more interesting.

Shade and water clarity also affected pond life: sunny ponds were better than shady ponds and clear water was better than murky water. Size of pond didn't seem to matter very much. We found more kinds of animals in larger ponds, but this seems to be mostly because we took more samples in bigger ponds. Larger ponds probably do contain more different animals, but over the range of pond sizes found in most gardens, the difference isn't enormous.

Like compost heaps, ponds had a noticeable effect on terrestrial garden wildlife. We caught more adult hoverflies and craneflies in gardens with ponds than in gardens without ponds. The larvae of most hoverflies eat aphids, but the larvae of a few very common species are aquatic, including the drone fly, named for the remarkable resemblance of the adults to honeybees. Similarly, although the larvae of the largest craneflies (daddy-long-legs), the familiar leatherjackets, are lawn dwellers, there are a lot of smaller species, many with aquatic larvae. Our Malaise traps caught so many flies that we weren't able to count them all, never mind identify them. However, if we had, it's likely that we would also have discovered more flies of other kinds in gardens with ponds, including a variety of midges and mosquitoes. A few of these flies bite people, but most of them don't, and all are important food for many other garden creatures, from birds to spiders.

> To sum up, it's obvious that both ponds and compost heaps do two important things. First, they provide habitats for those animals that spend all their time there and wouldn't manage to live in gardens at all otherwise. Second, they help to support all those insects that need special habitats (rotting vegetation or water) for their larvae, but live elsewhere in the garden as adults. No gardener with any ambition to attract wildlife should be without either.
>
> Pond size isn't very important, but the ideal wildlife pond should be in an open, sunny spot, and contain plenty of plants and no fish.

GARDEN DOCTOR: SELLING YOUR GARDEN TO WILDLIFE

WHEN YOU COME TO SELL YOUR HOUSE, YOU KNOW THERE are many things you can't do anything about. The size of the house, number of rooms and distance to the nearest school and shops are either outside your control or could be changed only with considerable difficulty and expense. On the other hand, there are many small things you could change very quickly, and some of these may have a disproportionate effect on its attractiveness to potential buyers. Some new paint, cleaning the bathroom grouting, a few fresh flowers. Even hiding some of those hideous souvenirs from Torremolinos might pay dividends.

In much the same way, the basic features of your garden may be more or less fixed, and in any case you may like your garden the way it is. Despite my advice to plant more trees and leave some long grass (OK, I haven't asked you to do this yet, but I soon will), your devotion to garden wildlife may not run that far. So, without altering the fundamental fabric of your garden, are there any relatively quick and easy (and preferably cheap) things you can do to benefit wildlife?

There's no shortage of suggestions in books, magazines and television programmes, but do any of them actually work? And when I say work, I don't just mean work well enough to satisfy the statistical demands of scientific proof. I mean do they work so that any gardener trying them will see a return on his or her effort most of the time? And if they cost anything, are they worth the money? To try to answer this question, the BUGS project set out to test a few.

There was a limit to what we could do to other people's gardens. First, anything we did had to be easily undone at the end of the study. Second, our volunteer gardeners had to be able (and willing) to help us. This wasn't always the case. For example, we wanted to see what happened if patches of lawn were left to grow long for a season or two. Why did we want to do this? Well, long grass should provide many benefits for wildlife. A recent study of grassland bugs found ten times more species and fifty times more individuals in long grass compared to grass mown every week. It's also interesting that Jennifer Owen's garden, while an admirable wildlife garden in most respects, failed to arouse the maternal instincts of many butterflies. Only the whites laid eggs in Owen's garden, mostly on her cabbages. Meadow browns, for example, are relatively common garden butterflies but were never persuaded to breed in Owen's garden, which did not have any long grass. Meadow brown caterpillars eat grasses, but unsurprisingly they are not attracted by lawns. Given some long grass, however, meadow browns are very happy to breed in gardens – they do so every year in the long grass in my garden.

Bumblebees didn't nest in Owen's garden either. By far the commonest bumblebee in her garden, and in our Sheffield gardens, was *Bombus pascuorum*, a smallish, rather tawny, slightly scruffy bee. *B. pascuorum* usually nests in long grass, at the base of grass tussocks, eventually creating a small mound of moss and grass stems. *B. pascuorum* nests every year in the long grass in my garden. There's nothing special about my garden, so it seems likely that bumblebees and meadow browns (and perhaps other grass-feeding butterflies) might be persuaded to breed in most gardens, as long as long grass is available. Unfortunately, this was an idea that was destined to remain untested, since not enough of our volunteers were happy to go along with the idea of allowing some grass to grow long. Most gardeners are proud of their lawns and, despite recent research showing that many would like to spend less time on lawn care, seem to be simply unable to resist the lure of the lawnmower.

> *Long grass is good for wildlife, and in short supply in gardens. If you want to leave some long grass, while at the same time convincing the neighbours that you are not some kind of dope-smoking layabout, by all means introduce wild flowers into the grass and call it a wildflower meadow. Most wildlife, however, will take no notice of the flowers — it's interested in the undisturbed long grass, not the flowers.*

Another bright idea was to see if nettle-feeding butterflies could be persuaded to breed in our gardens. We already suspected it might be hard to get this to work, so we hit on the cunning plan of luring adult butterflies into gardens with tubs of buddleja and then persuading them to breed by providing tubs of nettles next door. However, a proper scientific test of this idea depends on having some gardens with nettles and buddleja and others with just nettles on their own. Here the problem was not the willingness of our gardeners to help, but their ability. Plenty of gardeners were happy to play host to tubs of nettles, but examination of their gardens revealed that almost all of them already grew buddleja. Make a note: no need to advise wildlife gardeners to grow buddleja – they already do.

NETTLES

Nevertheless, the BUGS project was able to look at nettles in isolation, since very few of our gardeners already grew them. Peacocks, red admirals, small tortoiseshells and commas are among the commonest and most attractive of garden butter-flies, and the caterpillars of all of them feed exclusively on nettles. Not surprisingly, therefore, every wildlife gardening book ever written recommends growing nettles to encourage butterflies to breed in the garden, despite the absence of any evidence at all that this actually works.

We raided the nettle patch in my garden (which has never been quite the same since) for pieces of rhizome, which we planted in 20 litre tubs containing a mixture of peat-free compost, horticultural sand and fertilizer. We then placed one of these tubs in a convenient border in twenty of our gardens in June 2000. We went back at intervals throughout the rest of 2000, 2001 and 2002 to check for butterfly caterpillars. To see if the presence of fresh new growth attracts butterflies, we cut half the stems down in the summer of each year. Our nettles grew well, but to be honest, if I were a butterfly I wouldn't be very impressed by one tub of nettles. So in 2002, we added an extra three tubs to half the test gardens to create large 'four-tub' nettle patches. Once again the extra nettles came from the long-suffering patch in my garden, which had partially re-covered by then.

By now, I'm sure you can hardly contain your impatience to know if our nettles were overrun by butterfly caterpillars, so I'll quickly put you out of your misery. Sadly, reader, they were not. Over the three years of the experiment, we found two caterpillars of the comma butterfly on a single patch in 2000. Increasing the size of patches in 2002 made no differ-ence to the outcome – all our four-tub patches were completely untouched.

Perhaps we shouldn't have been surprised. Three nettle patches in Jennifer Owen's garden remained unused for fifteen years, even though adult butterflies were abundant in her garden. What's more, no butterfly has ever shown any interest in my nettle patch, even before it was reduced to a pale shadow of its former self. In fact, there are at least two good reasons why butterflies may not use nettles in gardens. First, nettles are hardly in short supply – indeed they have a claim to be Britain's commonest wild plant. Moreover, while our patches of nettles were larger than many garden owners might like (and many of our gardeners wouldn't have nettles in their gardens at all), maybe they still weren't big enough.

Our nettle patches weren't completely useless. They did attract other much less conspicuous nettle-feeding insects, including two moths. However, given the great abundance of nettles generally, it's extremely unlikely that garden nettles make any real contribution to the sum of urban biodiversity.

> *In the biggest relief since Mafeking, wildlife gardeners need feel under no obligation to grow nettles. On the other hand, nettles are just as good as comfrey as a nutrient-rich mulch or fertilizer, so there are other reasons to grow them. And if you do, perhaps your local butterflies will take a fancy to them. Just don't count on it.*
>
> *I also strongly suspect that growing thistles for the caterpillars of painted lady butterflies would be equally futile. And although thistles are an excellent source of nectar, so are many other more attractive plants.*

NESTS FOR BUMBLEBEES

Think of garden wildlife and you think of bees. Along with birds, bees are the main contributors to the sights and (especially) the sounds of the summer garden. All bees feed throughout their lives on nectar and pollen, so the most obvious thing you can do for them is to grow as many flowers as possible over as long a period as possible. No problem there – the interests of bees and gardeners coincide exactly. But bees need nests too, and to understand exactly what that means, you need to know that bees come in three different varieties. First, social bees, which build large communal nests, with a queen who lays the eggs and female workers that collect the pollen and nectar. Uniquely among social bees, honeybees make nests (hives) that persist from one year to the next. To get them through the winter, they store honey, a habit that makes them very useful to us. Honeybees are common in gardens, and the activities of beekeepers ensure that they hardly need any extra help from gardeners.

Bumblebees are also social, but their nests are strictly annual affairs, abandoned at the end of the summer. Only mated queens survive the winter, setting out to establish new nests in the spring. There are eighteen or nineteen species of true bumblebee in Britain (a few other cuckoo bumblebees sponge on them, as their name suggests). Six were common and widespread in our Sheffield gardens. Four of these look rather similar, with variations on the familiar stripy yellow, black and white jerseys. A fifth (*Bombus lapidarius*) is black with a red bottom, while the sixth and commonest (*B. pascuorum*) is slightly smaller, and a rather tawny colour all over. All are conspicuously furry and friendly insects, stinging only if really provoked.

Queen bumblebees emerge from hibernation in March, April or May and start to look for nests – indeed, the enormous queens of *Bombus terrestris*, one of the earliest species, are a familiar sign that spring is not so far away. But not all bumblebees are looking for the same thing. Of the six common species, three usually nest underground, often in an old mouse hole, while the others nest above ground. I've already mentioned that *B. pascuorum*, the commonest garden bumblebee, nests every year in the long grass in my garden. The hole-nesters in particular are quite adaptable, and have been found nesting in rolled-up carpet, in or under garden sheds, under an upturned sink, in bird boxes and compost heaps, in an old armchair, a cushion, among water-pipe lagging and even in a heap of coal.

Do bumblebees nest in your garden? Recent research suggests that bumblebees quite often nest in gardens, but that their nests usually escape attention. It's hopeless to attempt to spot the nest entrance itself – what you need to see is the bee traffic, the workers entering and leaving the nest to collect pollen and nectar. Because bumblebee nests contain only 200–300 bees (unlike honeybee hives, which have around 30,000 bees), this traffic is quite sparse and easily missed.

The secret is first to make a note of the likely nesting sites in your garden. Hedge bases, holes beneath sheds or garages, unmortared retaining walls and patches of long grass are all suitable. Choose a warm, sunny day, get a cup of tea, make yourself comfortable, and watch your possible nest sites for at least twenty minutes each. If there's a nest, you should see a few bees come and go during this time.

If you want to identify your bees and learn more about their ecology and behaviour, I recommend Bumblebees *by Oliver E. Prŷs-Jones and Sarah A. Corbet (Richmond Publishing Company, 1991).*

In an attempt to persuade bumblebees to nest in our gardens, we tried two nest designs. First, terracotta pots, with the drainage hole sealed, upturned on a tile sitting on a brick, so that the lip of the pot overhung the tile by about an inch. We also tried the same pots, but this time with the drainage hole unsealed, placed on a tile and buried just below the ground surface so that the drainage hole was the sole access to the pot.

You may have seen nests designed to appeal to hole-nesting

bumblebees in gardening catalogues, but they aren't cheap, and we certainly couldn't afford to buy enough to carry out a proper trial. So we took a standard design from a bumblebee expert and took it to a local workshop for people with learning difficulties. The nests were only a small departure from their normal output of bird boxes, so they were happy to make us thirty sturdy boxes from untreated exterior plywood. Any bee, I thought when I surveyed the finished product, would be proud to nest in such a snug and desirable residence.

All nests were supplied with a starter pack of bedding material (a handful of upholsterer's cotton), and then spread around our test gardens, in sheltered locations out of direct sunlight, for example at the rear of a flower border or at the base of a hedge or shrub. We did this in three separate years. In the last year, in a final effort to make our nests irresistible, we also tried some of our upturned pots and wooden boxes with bedding that had been used by domestic mice (courtesy of a local pet shop). If hole-nesting bumblebees that are commonly observed to nest in old mouse holes detect these by smell, we hoped that used mouse bedding would help searching queens to find our nests.

After all that effort, it would be nice, would it not, to be able to report some success. Sadly, I have to tell you that not one of our nests was used, in any garden, in any of the three years. Our upturned, buried pots were quite popular with hibernating newts, and spiders everywhere seemed to be grateful for them, but bees? No. This is such a sad tale that I'm glad there is a (slightly) hopeful postscript. Recent research shows that queens of hole-nesting bumblebees are pretty choosy about where they even *look* for nest sites. With bumblebees, it really is location, location, location. Since they seem to prefer old mouse holes, they look in the sorts of places they expect to find them. In practice, this means that they search along hedge banks, rather than in open, flat ground. Now, Sheffield being the hilly place it is, many gardens slope quite steeply, and mine is no exception. There are terraces in my garden, some edged with low dry stone walls, and I see bumblebee queens searching along these banks for nest sites every spring. Actual nesting is unpredictable, but at least one hole-nesting species nests in my garden in most years. Since it's clear that bumblebees are at least looking for nests in my garden, I tried three of our plywood boxes in my garden in 2003. All were carefully sited in places I had seen queens nosing around in previous years. To my surprise and relief (joy would not be too strong a word), one of them was occupied by a bumblebee nest. All that woodwork wasn't completely wasted after all. Nevertheless, honesty compels me to report that the same three nests, in exactly the same places, remained quite unused in 2004, 2005 and 2006, so it seems that chances of failure are high.

Persuading bumblebees to nest in your garden is possible, but like any good estate agent, you need to know what your customers are looking for. My experience confirms that you need not only the right nest but also the right location, and even then you need to be lucky. If your garden is quite flat, this probably means creating some kind of earth bank. The soil excavated from a pond is a good source of material.

A bought nest, or one you made yourself, can be placed on this bank or even incorporated in it. Alternatively, you could edge the bank with large stones or logs, to make some inviting holes to encourage a queen to investigate. This is at least as likely to work as installing an artificial nest.

Research also shows that queens of surface-nesting bumblebees are attracted by tussocky grass. This isn't something you can create overnight, so if you let some grass grow long in the hope of attracting bumblebees, be prepared to be patient.

NESTS FOR SOLITARY BEES

Honeybees and bumblebees are social, but there are many more species of solitary bees. As their name suggests, solitary bees don't make social nests with queens and workers. They just make a few cells in a convenient hole and lay an egg in each one. The nest is provisioned with pollen and nectar, but there is normally no parental care. By the time the new generation of bees emerges in the following year, the parent insects are long dead. Nest sites are very various – hollow plant stems, dead wood, holes in the ground, masonry. Some species, known as leaf-cutter bees, make their nest cells from pieces of leaf and are responsible for the holes cut out of the edges of your rose leaves. There are many species and all are excellent pollinators and completely harmless, so it pays to make an effort to attract them. The larger solitary bees look a bit like small bumblebees, but most species are smaller and much less conspicuous.

Recently, a small industry has grown up supplying gardeners with nests for solitary bees, but these nests were too expensive for us to use in large numbers, so we made our own:

- Clean, used tin cans, painted green on the outside and packed with paper straws.
- Blind holes drilled in untreated sawn 'red pine' blocks (50 × 50 millimetres in section, 100 millimetres in length). Different blocks had holes of different diameters: 10, 8, 6 and 4 millimetres.
- Lengths of bamboo cane packed into plastic drainage pipes (coloured terracotta, 110 millimetres in diameter, cut in 200 millimetre lengths) and capped at one end with a 112 millimetre postal tube bung.

All three nest designs were hung from a 1.5 metre stake in sunny, south-facing locations in twenty gardens in late March or early April. After our disastrous attempts to persuade bumblebees to nest in our gardens, you'll be relieved to hear that solitary bees were much easier to please. Over three years, our nests were used in all twenty test gardens, although not in every garden in every year, and the actual number of nest holes occupied was usually low. Bamboo sections and 4 millimetre holes in wooden blocks were the most popular, but all designs were used in some gardens. By autumn, occupied nest holes could easily be spotted by the entrance seals of mud or saliva and wood fibres. We put occupied nests in an unheated building until late spring, and then moved them to a laboratory where the new generation of adult insects were collected and identified during the course of the summer in emergence traps.

The big surprise was the variety of insects that used the nests. Only two species of solitary bee emerged, but nine species of solitary wasps used our nests. If the word 'wasp' makes you run a mile or reach for a rolled-up newspaper, don't worry – solitary wasps are much smaller than their social cousins and quite harmless to humans. On the other hand, they're not harmless at all to a whole range of insect pests in your garden. They catch and paralyse aphids, caterpillars, flies and leafhoppers and use these to provision their nest cells. They are therefore among the best pest controllers in your garden and definitely to be welcomed and encouraged. Finally, our nests were also used by three species of ruby-tailed wasps – scroungers that let solitary wasps and bees do all the hard work and then lay their eggs in their nests. When the larva emerges, it eats the larva of the luckless wasp or bee and sometimes (to add insult to injury) its stored food supply as well. Unpleasant as this lifestyle may be, the insects themselves are exceptionally beautiful, usually coloured in various shades of metallic red, green or blue.

Encouraged by the success of this trial, in 2002 we carried out an additional, simpler but much larger experiment. We put two wooden blocks containing all four hole diameters (10, 8, 6 and 4 millimetres) in each of eighty gardens. Our gardeners fixed the blocks to any convenient support, usually a wall, fence or tree, one in a sunny location and the other in the shade. Again they worked well, attracting nesting bees or wasps in almost half the gardens tested, but the blocks in sun were much more popular than those in shade.

> *Solitary bees and wasps are an important part of the biodiversity of the average garden and, unlike some wildlife, there's no downside; for both pollination and pest control, you want as many as possible of both. The garden managed with wildlife in mind will already provide many suitable nest sites – dead wood, plant stems, old walls – but our experiments show that many species will readily use simple nests made from cheap, widely available materials. So what are you waiting for?*

DEAD WOOD

Looking at today's tidy countryside, it's hard to imagine a landscape almost completely covered by trees. It's also hard to comprehend just how much dead wood there would have been lying (and standing) around only 5,000 years ago, when Britain was one enormous forest, compared to how little there is now. Not surprisingly, animals that need dead wood (including a huge number of beetles) are among the most threatened species in modern Britain. More than 1,700 British insects are reckoned to depend on dead wood, and at least 300 of them are endangered. Of invertebrates that depend on dead wood, 17 per cent have not been seen in Britain since 1900 and must be presumed extinct here. Around three-quarters of endangered woodland beetles have larvae that live in dead wood, mostly eating the wood itself, but sometimes preying on other animals that also live there. Some of these beetles are now so rare that they are confined to a single ancient wood or park, such as Windsor Great Park or Epping Forest, where there has been a continuous supply of dead wood since prehistoric times. The iconic dead wood species is the stag beetle, one of Britain's largest insects and, despite its fearsome appearance, quite harmless to humans.

Although the key to the survival of most of these wonderful animals is better management of the woods and parks where they already occur, gardeners can certainly make a contribution, particularly to the conservation of the more widespread species. Stag beetles are declining in the wild, but still not uncommon in gardens. Unfortunately we couldn't expect to attract stag beetles to our gardens, firstly because they only occur in the south, and secondly because they prefer decaying tree stumps, a hard habitat to create overnight. Nevertheless, we did our best.

We stacked freshly cut logs of silver birch in each of twenty gardens, six logs per garden (lengths 600–800 millimetres, diameters 70–150 millimetres), in November 2000. Log piles were located out of direct sunshine, at the base of vegetation or at the rear of borders. Two years later, in October 2002, we returned to see what we'd caught. We first noted anything living under the logs, then took the logs and the litter that had accumulated around them back to the lab for a closer look. We didn't expect to find anything very exciting, and we weren't disappointed – our logs were too small, and two years isn't long enough for them to rot down very far. Nevertheless, our log piles were already home to several fungi and a wide range of animals, including frogs, mites, harvestmen, spiders, centipedes, millipedes, woodlice, earthworms, flatworms, snails, slugs, springtails, ground beetles, rove beetles, bugs and flies. Most of these are harmless, and even if you think you already have enough of some of them (slugs, for example), don't be put off providing dead wood as at least the slugs were accompanied by animals that eat them (frogs and predatory beetles).

Many insects are quite choosy about the kind of dead wood they prefer – still part of a tree, or lying on the ground. Leave dead branches attached to the tree unless they are actually dangerous (if you're not sure, get professional advice). If you have to remove a tree, one of the best things you can do for Britain's threatened wildlife is to leave a substantial stump – the taller the better. In the wild, hollow trees tend to fill up with bird droppings and dead leaves. To provide a passable imitation of this in the garden, try chopping some holes in the top and filling them with leaf mould and chicken manure.

If you don't have a convenient dead tree, the next best thing is a pile of logs, the bigger the better – both the pile and the individual logs. The logs shouldn't cost anything – just keep your eyes open and, sooner or later, there'll be someone felling a tree or clearing up a fallen tree. Stack the logs somewhere shady: most dead wood species like it damp. Partly burying the lowest logs will keep them damp and speed up decay, which is good for wildlife – many beetles like wood that has begun to rot. Even the largest log pile will eventually rot and collapse, but don't try to move or dismantle it, just pile some more logs on top.

Even if you don't have a tree stump or any logs, you can try a technique recommended by the People's Trust for Endangered Species. First make lots of large holes (at least 30 millimetres diameter) in the sides and bottom of a plastic bucket. Put some large stones in the bottom, and then fill it with a mixture of one-quarter garden soil and three-quarters woodchips (hardwood preferably, but softwood will do), and bury it so that the top is flush with the ground. As the contents rot, top up with more soil/woodchip mixture. This device is designed specifically to attract stag beetles, but it should also attract other insects that breed in dead wood. If stag beetles use your bucket, or if you just see them around (and especially if you live in northern England or Scotland), the trust (www.ptes.org) would like to hear from you.

PONDS

We had looked at real garden ponds, which I talked about in the last chapter, but, being scientists, we couldn't resist setting up some small experimental ponds in gardens that lacked them. To some extent, these were just an interesting academic exercise, but they did have two serious objectives. First, just how small can a pond be and still attract some wildlife? And second, what sort of wildlife can you expect to colonize a small pond, without any human assistance?

We bought nineteen plastic planter troughs (length 700 millimetres, width 300 millimetres, depth 250 millimetres), each providing a volume of 28 litres and a water surface area of just under a quarter of a square metre. Clearly, our volunteer gardeners didn't want us digging holes in their gardens, so the troughs were not sunk into the ground. So that animals such as frogs could get in and out, we fixed a piece of stiff plastic mesh as a ramp from the ground up and into the pond. We put a layer of horticultural grit on the floor of each pond, and filled it with tap water. To make sure that the first arrivals didn't simply starve to death, we added four pieces of Canadian pondweed (*Elodea canadensis*) and 300 water fleas (*Daphnia*) to each pond. We also hoped that the water fleas would eat any algae and stop the water turning green.

The ponds were set up in July 2000, in positions that didn't receive sunshine throughout the day in summer, so as to discourage algal growth, but were not covered to prevent leaves falling into them in autumn. We had a look at them in August 2000 and July 2001 to see how they were getting on, but we didn't attempt a full inventory of each pond until June 2002.

Despite their small size, the nineteen ponds did rather well. The pondweed grew well and the water fleas thrived in most of them. Filamentous algae was a problem in most ponds early on, but it was much reduced by June 2002. Midges and mosquitoes were the first arrivals – more than half the ponds had eggs of both within ten days of establishment, and after one month all but two ponds had been colonized. Mayfly nymphs, water beetles and the aquatic larvae of hoverflies and craneflies were all found in a few ponds. Snails, water slaters (*Asellus* spp.) and ostracods (tiny Crustaceans) were frequent – presumably eggs of all three came in with the pondweed. Disappointingly, few other aquatic insects with flying adults colonized the ponds – we never saw familiar pond inhabitants such as water boatmen and pond skaters. Amphibians didn't breed in any of our ponds in either of the two available seasons, but juvenile and adult frogs were recorded in seven ponds, despite the only access being a strip of plastic mesh.

Our survey of real garden ponds suggested size isn't very important, and our experimental ponds supported this view. Whatever the lower limit for the size of a garden pond may be, our ponds (only the size of a window box) were obviously above it. Despite probably freezing solid in winter, most ponds developed thriving animal communities and there were no serious problems with algae. Plenty of animals turned up and, if you started out with a bucket of water from an established pond, there would no doubt be many more. The lesson is that there's probably no garden too small to have its very own pond.

CONCLUSION

The broad conclusion is that while some simple methods for increasing the biodiversity of gardens may be very effective, others are very unlikely to work. Nor does this have much to do with how much they cost or with how often they are recommended. Nettles for butterflies are a waste of time. So are bumblebee nests, even at a tenner each, which is what ours cost, and certainly at £30 each for the commercial version.

On the other hand, dead wood is guaranteed to work, and your garden is full of small animals looking for a hole to nest in, so providing almost any natural material with holes in will pay dividends.

It's a jungle
out there

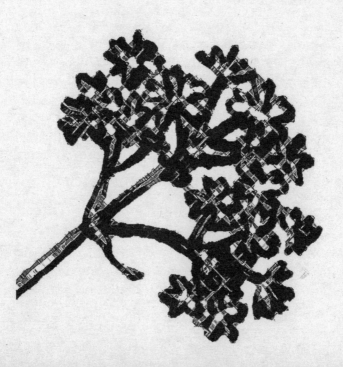

I'VE JUST BEEN OUT TO LOOK AT MY HOSTAS AND I HAVE TO say they look terrible, like the morning after a lace-making festival. However, this level of damage is exceptional. Few other things in my garden look even half as bad, apart from my Solomon's seal, which has been completely consumed by some horrible caterpillar. When you come to think about it, this relative lack of damage is surprising. Plants just sit there all summer, looking like so many plates of salad, yet by the end of the growing season, most of them remain largely uneaten. Ecologists noticed this long ago and coined the phrase 'The world is green'. More often this was phrased as a question: 'Why is the world green?', since it's not immediately obvious why most of that tempting green stuff should survive the onslaught of all those beetles, caterpillars and slugs. Not to mention larger animals like rabbits, voles and deer.

One attractive possibility is that most plant material is better defended than it looks. Many plants have gone to considerable lengths to discourage herbivores. Sometimes this is obvious, in the form of thorns, spines and hairs, but more often it's much less obvious, in the form of various toxic chemicals. Unfortunately, these poisons are only a small part (at most) of why plants don't get eaten much more than they do. The problem is that given enough time, animals can usually come up with the answer to most plants' attempts at chemical warfare. They may even turn the poisons to their own advantage. The cabbage and its close relatives have tried hard to perfect the art of being inedible, and – depending on your opinion of Brussels sprouts – you may think they have succeeded. But the caterpillars of cabbage white butterflies haven't only learned to break down the toxic glucosinolates in cabbage leaves: they now use the smell of these chemicals as a foolproof way of locating their food plants. Ragwort is so poisonous that horses cannot be allowed to graze in fields where it grows, but cinnabar moth caterpillars now store the ragwort alkaloid in their own bodies as a defence against birds. Their orange and black jerseys advertise their borrowed toxins. At any one time, it's easy to think that particular plants have won (or lost) the war with herbivores. A ragwort that has lost every single square millimetre of leaf to cinnabar moths looks as if it's lost. On the other hand, bracken contains such a lethal cocktail of chemicals that hardly anything eats it, so it looks as if bracken has won. In fact, of course, this is just a snapshot of a single battle in a war that has already been going on for millions of years. Particular plants or herbivores may seem to be winning now, but things may well look quite different in another million years. The war between plants and the animals that eat them is a game in which both parties, like the Red Queen in *Through the Looking-Glass*, have to run just to stay in the same place, and the only prize is being allowed to carry on playing the game.

So if poisons don't protect plants from total defoliation, what does? The most important answer is that long before plant-eating animals have the chance to eat everything in sight, they attract the attentions of predators. In the short term, extra predators home in on places where there are plenty of herbivores to eat. In the slightly longer term, numbers of predators rise as a result of increased reproduction of animals already present. Even if herbivores initially get the upper hand, the predators always catch up in the end. Over most of the planet, most of the time, populations of plant eaters are kept at levels too low to do really serious damage to the plants they eat. Now and then outbreaks of some herbivores, such as the African migratory locust, escape their predators long enough to cause trouble, but these are rare exceptions to the general rule. Increasingly, of course, plant-eating insects get out of control because indiscriminate use of pesticides has wiped out their natural enemies, but that's another story.

Before I wander completely off the point, what does the control of herbivores by predation rather than by the food supply mean for gardeners? Well, I think that we would all agree that this state of affairs is good for both gardens and gardeners. In a world in which herbivores were limited by their food supply, and just kept on chewing until there was nothing left to eat, shredded hostas would be the least of our worries. But there's a downside. Because the fate of most plant-eating animals is to die young, there is a lot of carnage going on out there. Every day in your garden, hundreds or even thousands of creatures die grisly and often highly inventive deaths.

KILL OR BE KILLED

In the modern world, in which the chance of your being killed by a non-human predator is rare enough to be ignored as a serious hazard, attitudes to all this bloodshed are often rather ambivalent – when you watch a wildlife programme about the Serengeti, do you cheer for the lions or the zebras? For much of the time, gardeners are happy with the mounting body count, as long as it's the bad guys who are getting it in the neck. Gardening books, magazines and television programmes are full of advice on how to encourage predators like ladybirds and hoverflies, and generally try hard to promote the idea that predators are on our side in the war against pests. There is plenty of evidence to show that they are. Foresters know that wood ants are good for trees, because they eat tons of plant-eating bugs and caterpillars, and it's even worth transplanting the ants to new plantations. Research has shown that installing nest boxes in orchards to encourage great tits increases both the yield and quality of apples.

Unfortunately for wildlife gardeners, there are a couple of problems with this generally rosy scenario. One, which I've drawn attention to before, is that there has to be a balance. If you like ladybirds, you have to have some aphids. If you like blue tits, you have to have some caterpillars. You may not like midges and mosquitoes, but bats, swallows and house martins think they're delicious. If you want the wildlife you like, you have to have the wildlife you don't like. Second, despite the propaganda of the gardening media, it's not always the bad guys that get eaten, nor the good guys that do the eating. You may cheer on the spiders in your garden as they munch their way through wasps and flies, but you may be less pleased when they catch bees and butterflies. You may think fox cubs frolicking in your garden are charming, but all predators have to eat; flamingos used to live in the garden of Buckingham Palace, until the local foxes discovered them. I have to confess to mixed feelings last week as I watched a heron making off with a frog from my pond.

In an earlier chapter, I told you how hard it was to persuade bumblebees to use artificial nest boxes, and how pleased I was on the one occasion that I actually succeeded in doing so in my garden. A generally cheering story, you would probably agree, but unfortunately it doesn't have a happy ending. Venturing out into the garden one pleasant summer morning, I was astonished to find my occupied nest box upturned, the lid knocked off and the contents scattered around. A few worker bees were flying around, looking dazed and confused, but the nest was completely destroyed and there were no intact cells or larvae to be seen. So that was the end of my bumblebee nest, but who or what was the culprit? I don't know. I often see foxes in my garden, but I've no idea if they ever attack bumblebee nests. I wouldn't be surprised, since they are omnivorous and often eat beetles, worms and other small creatures. Badgers are certainly serious predators of bumblebee nests, but I've never seen any sign of badger activity in my garden, although they are occasionally seen in the neighbourhood. Badgers, indeed, are one of the acid tests of your commitment to wildlife gardening. Immensely strong and equally single-minded, badgers can do a huge amount of damage in a very short space of time. They like to eat earthworms, and if the earthworms happen to be under your lawn, well, it's hard luck for the lawn. They also eat hedgehogs and attack bumblebee nests – in fact one reason (although not the main one) why hedgehogs and bumblebees do so well in gardens is that they are relatively free there from the risk of predation by badgers.

I'm not trying to malign badgers, and I certainly don't want to give comfort to those who think they are awful vermin that should be exterminated. I'm simply making the point that the animals in your garden have their own lives and agendas, which may or may not coincide with your idea of the smooth running of a garden. A big part of successful wildlife gardening is a willingness to relinquish the level of control over your environment that you take for granted in the home, office and supermarket. Your garden may not be a particularly wild place, but the animals that live there don't know that. They really are *wild*life.

SPARROWHAWKS AND MAGPIES

Attempting to interfere directly in the lives of the animals in your garden rarely has exactly the intended effect, and in any case you need to know that the interference was justified in the first place. Take the example of declining songbird populations. During the last thirty years, populations of many garden songbirds have declined dramatically. The reasons are unclear, but for some gardeners, the cause is only too obvious – the increase over the same period in abundance of the sparrowhawk and the magpie, two major predators of song-birds. Since 1970, both the sparrowhawk and the magpie have spread eastwards in Britain, although the causes are very different. Sparrowhawks have spread into areas of intensive arable farming, following the decline in use of the organochlorine pesticides that had formerly eliminated them from these areas. Magpies, on the other hand, seem to have benefited from a fall in the numbers of lowland gamekeepers. At the same time, many familiar garden birds have declined, and all are taken as prey by sparrowhawks, while eggs or nestlings of many (but not all) are eaten by magpies. It looks like an open and shut case, but is it?

In fact detailed analysis of data from the British Trust for Ornithology shows conclusively that there is no connection between the decline in songbirds and the increase in sparrowhawks and magpies. The blame for songbird decline, and for many other undesirable changes in the British countryside over the same period, almost certainly lies with agricultural intensification. Sparrowhawks and magpies were just innocent bystanders. For gardeners, this is extremely good news. You can now enjoy the presence of magpies and (if you're fortunate enough to see them) sparrowhawks in your garden without feeling the need to reach for the shotgun. Of course they will eat some birds (predators have to eat), but that's life, and in any case, the numbers of birds taken by sparrowhawks is negligible compared to those killed by cats. What's more, sparrowhawks and magpies kill because they need to eat and feed their young, whereas cats just kill for fun. You may be reassured, in any case, that even top predators don't have things all their own way. An aged local fox recently chose my garden to die in, so I had the chance to have a look at his corpse. He turned out to be host to numerous ticks, each of them the size of a small grape, which I suspect didn't improve his temper. Of course they died with him, but that was their bad luck.

You sometimes hear hard-headed, no-nonsense types assert that 'the balance of nature' is a myth put about by social workers, hippies and other dangerous lefties, and that the only sure way to deal with pests is to hit them with everything in the chemical armoury. In reality it's this 'SAS-style' approach to gardening that is nonsense. The balance of nature is an objective scientific fact, and its usual outcome is that plants survive the onslaught of herbivores without suffering life-threatening damage, give or take the odd hosta.

Nevertheless, let me mention two important caveats. First, for most of the time your garden is no more in perfect equilibrium than, say, the economy or the weather. In any one place, at any one time, plant eaters may temporarily escape control by their predators. And even if the herbivores are under control, that's not the same thing as eradicated, and the survivors do have to eat. So we gardeners who really care about wildlife must learn to tolerate holes in our hostas, aphids on our roses and the occasional grub in our raspberries.

Second, predators will only control herbivores if they are given a fighting chance. If we have effectively sabotaged the whole system, for example by introducing a whole bunch of extra herbivores (such as rabbits and several different kinds of deer) and exterminating the predators (wolves, bears, lynx) that would have kept them under control, you can't expect the balance of nature to function effectively. If your garden is decimated by rabbits or muntjac, don't expect any help from Mother Nature any time soon.

A HELPING HAND

Having said all that, there's no harm in some gentle, non-chemical intervention if you think things are getting a little out of hand. If the buds on your rose bushes are invisible for aphids, squashing a few or blasting them off with a hosepipe won't do too much damage to your organic credentials. If cabbage white caterpillars are turning your sprouts into confetti, picking them off is perfectly OK. Don't waste them – put them on the bird table, where robins or blackbirds can enjoy them. But if you're going to adopt this hands-on approach, make sure that you know who your enemies really are, and how to tell them from your friends. Predators have enough problems without having to suffer friendly fire from gardeners.

Sometimes a failure of taxonomy isn't disastrous, or even very important. If you can't tell millipedes from centipedes, don't worry. Not only are you in good company, but it really doesn't matter – both are useful animals in the garden. Ground and rove beetles are also quite easy to spot and both need encouraging – ground beetles are a major consumer of slugs. Another beetle we all recognize is the ladybird, or at least we think we do. As well as *the* ladybird (the seven-spot), there's the equally common two-spot, which is very variable – normally red with two black spots, but quite often red with various black patterns or black with red spots. There are many other ladybirds, with ten, fourteen and even twenty-two spots, but most share the red or yellow and black colouring, which advertises their nasty taste to birds that might be tempted to eat them. Ladybird larvae are equally voracious consumers of aphids, so make sure that you can recognize them. The larvae of our commonest ladybirds are blue-black with a few conspicuous yellow or white spots. Most beetle larvae (and larvae in general) have short, stubby legs, or no legs at all, but ladybird larvae have long legs and forage actively during the day. Like ladybirds, lacewings are simple enough to recognize, but their larvae are not quite so easy. They are rather similar in shape to those of ladybirds, but most are coloured a dull brown, mottled with darker spots and markings. Some species festoon themselves with the remains of their victims, eventually looking more like piles of walking debris than anything else.

Hoverflies get my vote as 'gardeners' favourite insect'; the adults of most species are colourful, harmless pollinators, while the larvae of many eat aphids. If you were designing an insect, you could hardly ask for more. However, hoverflies have a trick that has stood them in good stead for millions of years but sometimes gets them in trouble with gardeners: they don't sting, but they look a lot like insects that do. Many insects avoid being eaten by predators (especially those, such as birds, that hunt by sight rather than scent) by tasting nasty and advertising the fact with bright warning colours. Some variation on black and red or black and yellow is usual, and birds soon learn to avoid insects with these combinations of colours. Bees and wasps are classic examples. Actually bees taste OK, but they are tough to eat and they sting, so birds generally avoid them. Wasps, on the other hand, sting *and* taste really horrible. Hoverflies taste nice, but have hit on the clever idea of looking like bees and wasps. The curious thing is that they don't do this very well. Although they fool some novice gardeners, once you get your eye in, you soon realize that most common hoverflies don't look much like wasps or bees at all. This certainly isn't because they can't – if natural selection can make one caterpillar look like a twig, and another exactly like a bird dropping, making a fly look like a wasp or a bee is child's play. In fact one uncommon hoverfly looks so much like a bumblebee that its own mother probably couldn't tell them apart.

No one is quite sure why most common hoverflies mimic wasps rather poorly, but one possible reason is that there are actually many different kinds of wasp. As well as the familiar large social wasps that cause so much trouble at barbecues, there are many other solitary wasps of various sizes, shapes and patterns of black, yellow and red. All are bad to eat, and it looks as if birds soon learn to avoid any wasp-like insect. So although hoverflies could do a perfect replica of a particular wasp, it looks as if the most effective strategy in the long term is to do their best to try to look like some kind of generalized *über*-wasp.

Don't mistake hoverflies for wasps, although wildlife gardeners have no reason to dislike either. The key is that hoverflies, like all flies, have short antennae. Wasps, and their relatives the bees and ants, all have long antennae.

You are perhaps more likely to kill hoverfly larvae by mistake, which is a serious error, to be avoided at all costs. Hoverfly larvae are flattened, legless and maggot-like. In fact they look more like lumps of animated snot than anything else. Not very handsome animals at all, but console yourself that they look far worse if you're an aphid.

To help you identify the insects mentioned, plus plenty of others, I strongly recommend Collins Guide to the Insects of Britain and Western Europe *(Collins, 1986) by Michael Chinery. This is currently out of print but should by now have been replaced by the similar* Insects of Britain and Western Europe *(A&C Black, 2007) by the same author.*

COMPOST AGAIN

This chapter began by remarking how most greenery seems to make it to the end of the growing season without being completely consumed. A logical consequence of this is that the fate of most leaves is to fall to the ground at the end of the growing season. If you spend a lot of time raking them off the lawn, or trying to stop them falling into the pond, you will hardly need this pointing out. But you will also have noticed that fallen leaves, and dead vegetation in general, do not usually accumulate from one year to the next. Woodlands are not knee-deep in dead leaves. However much compost you make and shovel on to your garden, there's not much sign of it a year later.

What this means is that while the numbers of animals that eat living plants may be limited by their natural enemies, the animals (and other organisms) that consume dead plants are limited by their food supply. The numbers of soil micro-organisms, fungi, earthworms, millipedes, woodlice and springtails – the organisms that recycle everything in your garden, and on which its fertility depends – are limited by the supply of organic material. It's easy to forget that soil is a living community, and the size and activity of this community depends on its food supply. A poor garden soil may contain only thirty earthworms per square metre, while a fertile soil with plenty of organic matter may contain ten times as many.

At the risk of repeating myself, I cannot over-emphasize the importance of soil organic matter. Because soil biodiversity is both invisible and poorly understood, it's easy to ignore it, but you do so at your peril. Even if all you care about is growing better cabbages, it's worth knowing that an annual dose of leaf mould doubled yields from vegetable plots at the Henry Doubleday Research Association. But that extra growth above ground is merely a symptom of the enormously increased life and activity out of sight below ground.

Maintaining soil carbon is easy: make as much compost as you can, grow lots of plants and go easy on the digging. And if you care about global warming (as you should), reflect that in the UK, plants contain only just over 1 per cent of our total national store of organic carbon – the rest is in the soil.

A CHAPTER
ABOUT BIRDS

THE SMALLER CREATURES IN YOUR GARDEN ARE NOT ONLY more numerous than the larger ones: they are also more important. If all the birds in your garden disappeared tomorrow, a year from now it wouldn't look very different from how it does today. On the other hand, if all the earthworms and bacteria disappeared, your garden would cease to function in a matter of weeks, if not days. Nevertheless, birds *are* important for at least two reasons. First, a healthy bird population is the most visible sign that all the other vital but less-than-obvious wildlife in your garden is working well. If your worms and bacteria are all present and correct, the birds probably will be too. Second, birds are charming, interesting and beautiful. Gardening has few pleasures to compare with watching the antics of a party of greenfinches on a bird feeder, and certainly none that can be enjoyed so reliably from the warmth and comfort of an armchair in the middle of winter. It's also worth noting that recent research has revealed that gardens are much more important for birds than anyone suspected – for some species, there are as many birds in gardens as in all the rest of the country put together. Anyway, what's the point of a book on wildlife gardening that has nothing to say about birds?

For numerous reasons, the BUGS project did not attempt to study garden birds. Nevertheless, plenty of others have. As an object of scientific study, garden birds have one overwhelming advantage: there aren't very many different kinds. Most attempts to involve garden owners, and amateurs generally, in the study of biodiversity in gardens run into one insurmountable obstacle, which is the sheer diversity of small creatures that live in gardens. With the possible exception of butterflies, studying garden invertebrates is a job for a relatively few dedicated individuals. On the other hand, garden birds are big, easily identified (with very few exceptions) and strictly limited in number. The average gardener might see twenty different species, and even an exceptionally diverse garden is unlikely to be visited by more than forty.

Not only are there relatively few garden birds, but there is a huge infrastructure in Britain dedicated to helping gardeners attract and identify them. This enthusiasm has been exploited by the British Trust for Ornithology (BTO), which has run the Garden BirdWatch scheme since 1995. Over 16,000 gardeners now report the birds that visit their gardens, allowing the BTO not only to discover which birds visit gardens, but also to begin to reveal the long-term trends in their abundance. Much of the information in the rest of this chapter (although not all) is taken from the published findings of the BTO survey.

GARDEN SIZE AND LOCATION

Before looking at the results of the BTO survey, it's worth pointing out that there is a geographical trend in bird diversity across Britain. There simply are many more different kinds of birds in south-east England than in Scotland. Gardeners in Canterbury or Guildford will see more return on their efforts to attract garden birds than will gardeners in Stornoway, and there is nothing any of us can do about that. It will be a long time before Scottish gardeners can expect to see ring-necked parakeets, which have established a large breeding population in the Home Counties and the Thames valley. Nor does this imbalance apply only to birds. The distribution of British wildlife is recorded using the 10 × 10 kilometre squares of the national grid, and the 'best' square for plants is SY98, which has Corfe Castle in the middle and Wareham in the top left corner. This square has 844 native wild plants, while the best Scottish square (NT67 near Edinburgh, with Dunbar in the corner) has just 619 species. SY98 is also one of the best squares in Britain for butterflies, although the very best square is just a little further north, on Salisbury plain. If you wanted to attract as many different kinds of butterflies as possible to your garden, you wouldn't start by moving to Scotland.

As we discovered in an earlier chapter, no survey method is perfect, and all research data need to be approached with a healthy scepticism for the methods used to collect them. The BTO data are no exception. The clearest message from the BTO survey (and from other surveys of garden birds) is that you will see a greater variety of birds in large gardens than in small ones. Does this mean large gardens are actually *better* for birds? Well, maybe, but maybe not. Ask yourself this question: do the birds in your garden, flying in from the bird table next door or from a tree three houses away, actually look as if they either know or care where your garden begins and ends?

A more subtle point is that larger gardens are more likely to contain large trees and hedges and, as we'll see, these are good for most birds. So some of the apparent effect of garden size may be genuine (that is, big gardens *are* better), but the reason may have little to do with garden size itself. Another sign that birds are less than fascinated by garden size comes from the few species, such as collared dove, starling and house sparrow, that the BTO found were more likely to be seen in small gardens. These are birds that are strongly urban in their distribution, and urban gardens tend to be smaller than suburban and rural gardens. So just as most birds don't really favour large gardens, these species don't favour small ones – they just like to live in places where gardens, on average, are smaller.

It seems that birds, like most other garden inhabitants, are not much influenced by garden size. If you can provide the things that make gardens attractive to birds – and it will help a lot if at least some of your neighbours do the same – a small garden should be no barrier to attracting a wide range of birds.

The BTO data also show that garden location is more important than the features of an individual garden. Essentially a greater diversity of birds was seen in rural gardens; suburban gardens followed closely, with urban gardens coming a relatively poor third. Recall that this differs from the BUGS findings for garden invertebrates, which were unaffected by distance from the edge of the city (BUGS didn't look at rural gardens). This probably reflects a genuine difference between birds and most insects in the scale of their normal activities. Most insects breed and forage in just one or perhaps a few gardens. On the other hand, most birds range over much larger distances on a daily basis, and may also exploit quite different kinds of country at different times of year. Thus birds that live in farmland, moorland, broad-leaved woodland or conifer plantations are likely to turn up in rural gardens or suburban gardens on the edges of towns and cities. Many birds breed in rural habitats but will come to gardens in the winter to take food put out for them by gardeners. Clearly they are more likely to visit rural or suburban gardens to do this.

However, we should be cautious in accepting the apparent finding that garden location is *much* more important than features of individual gardens. Gardens in the BTO survey belong, by definition, to gardeners who are interested in birds, and are therefore not representative of British gardens as a whole. Even in a bird-mad country such as Britain, there are still plenty of gardeners who are not interested in birds. In particular we can assume that BTO gardens are towards the 'bird-friendly' end of the spectrum. It seems very likely that a genuinely random survey of garden birds (including some distinctly 'bird-unfriendly' gardens) would reveal a larger effect of what goes on within individual gardens.

> *For birds, location is important. If you live well inside the built-up area of a town or city, there are some birds you may never persuade to visit your garden. But don't imagine that you need to live in the country to see a wide range of birds. In her ordinary suburban garden in Leicester, Jennifer Owen managed to get her garden bird list up to forty-nine, although that massive total took fifteen years and depended on including birds such as the grey wagtail and skylark that were seen flying over the garden but never landed, and the cuckoo and green woodpecker, each seen just once. Including escaped cage birds and the odd racing pigeon would have increased the total to fifty-four species.*

The importance of garden location is one sign that birds have a life outside gardens. Another is that some birds may show up in your garden only at certain times of year. The most obvious examples are birds that visit Britain only in the winter, such as the fieldfare and redwing, or only in the summer, such as the swallow and whitethroat. In fact these seasonal visitors provide one of the best examples of how gardeners can transform the fortunes of some birds. Pick up any older book on British birds and you will find the blackcap described as a summer visitor, breeding in Britain but leaving for the Mediterranean in September. Recently, however, many blackcaps that breed in Germany or Switzerland have taken to spending their winters in Britain. It's clear why they might want to do this: birds that winter in Britain can get back to their breeding grounds up to two weeks earlier than the softies that spent the winter on the Costa del Sol, enabling them to grab the best territories and raise more young. One reason European blackcaps can do this is our recent milder winters, but another is the food provided by gardeners. Like most warblers, blackcaps are mainly insect eaters, but in winter they are happy to take bread, fruit and fat from bird tables. The result is that peak sightings of this 'summer visitor' in British gardens are now in late winter.

Even birds that live in Britain all year round may not be seen in gardens throughout the year. Sightings of many birds in gardens are at their lowest in autumn, a time when fruit and seeds are at their most plentiful in the countryside. Because many of the most common garden visitors show this pattern, many gardeners find that a lot of their birds have deserted them at this time of year. Patterns like this mean that numbers of birds in gardens are not always a good guide to how well birds are doing in the wider countryside. Beech mast is one of the favourite foods of coal tits, and in years with a good crop, this species may almost vanish from gardens in autumn. Siskins are another example of bird behaviour being influenced by gardeners. When tree seeds begin to run out in late winter, siskins may visit gardens in large numbers. The interesting thing is that siskins only discovered the possibilities offered by gardens in the late 1960s and their visits to gardens are still on the increase.

Long-term trends, on the other hand, are normally a better guide to the underlying health of the populations of garden visitors. For example, sparrowhawks were almost wiped out in many parts of eastern England by the effects of organochlorine insecticides, and before about 1970 were hardly ever seen in gardens. Since then the population has shown a steady recovery, reflected in increasing numbers of sightings in gardens. Interestingly, this recovery was initially confined to rural gardens, but suburban gardens have now largely caught up. In some other species, garden data show a long-term decline. Surprisingly, both the house sparrow and starling, two of our most abundant and most urban birds, have suffered significant declines over the last ten years. Even more intriguingly, both declines show striking regional differences. Starling numbers have fallen sharply in south-east England, but populations in Scotland, Ireland and Wales have remained stable. House sparrow numbers have also fallen in south-east England, but the decline has been far worse in Greater London. In both species the causes of the declines, and of the regional differences, remain unknown. Whatever the cause, however, gardeners have a particular responsibility to these mainly urban birds.

TREES AGAIN

As we've seen, garden location seemed to be more important for birds than the features of the individual garden, although the survey probably exaggerated the difference. Nevertheless, in those birds that showed a significant effect of individual garden features, this was always the same — more birds in gardens with trees and hedges, and fewer birds in more open gardens. Given the crucial importance of trees for invertebrate wildlife, and the importance of trees and hedges for nesting, this is hardly surprising.

> *If the welfare of beetles, flies and wasps isn't enough to persuade you of the merits of trees, perhaps birds will do the trick. The best single thing you can do for the wildlife in your garden is to find a young tree and leave it alone. Failing that, plant one.*

FOOD

The BTO Garden BirdWatch results are crystal clear about one thing: gardeners who provided food saw more birds than gardeners who didn't. Most members of the BTO survey, of course, have got the message and already feed the birds in their gardens. The rest of us could do more, and many of those who already provide food could use more imagination. Think of bird food, and most of us think of the standard peanut-filled bird feeder. Yet many birds don't like peanuts, many more don't like bird feeders and quite a few don't like either. Like most finches, chaffinches will eat peanuts, but they prefer to eat from the ground or from a bird table, so you won't see them on your bird feeder. Acrobatic goldfinches, on the other hand, are quite at home on bird feeders, but they are not fond of peanuts. Goldfinches have long, thin beaks, ideal for extracting the small seeds from teasels or thistle heads, so unsurprisingly they prefer smaller seeds. The best food for these birds is niger (or nyjer), the small seeds of *Guizotia abyssinica*, a tropical yellow daisy. Niger is an important oilseed crop in Ethiopia and some other African and Asian countries, so buying it will help not only goldfinches but poor farmers too. Some other birds, including siskins and redpolls, also seem to prefer niger seed.

If you're really serious about attracting birds to your garden, it pays to provide a diversity of foods and feeders. Some birds will take an irrational dislike to a particular design or even colour of feeder. For years I used the simplest, cheapest square peanut feeder, but was eventually forced to change by the depredations of my local squirrel population. The old feeder, painted red, was extremely popular with all kinds of birds, including tits, greenfinches and siskins. From the day I installed my new armour-plated, squirrel-proof peanut feeder, in a tasteful shade of pewter, greenfinches and siskins disappeared. Tits seemed moderately content with the new arrangements, but you could tell that even they weren't really happy. When I finally lost patience and bought an extra feeder designed to hold sunflower seeds, the first greenfinch showed up exactly fifteen minutes after I put it out, and my local greenfinches now eat about 2 litres of my sunflower seeds a week. In fact when a party of hungry greenfinches shows up, you can actually see the seed level going down, as if in a giant egg-timer.

> *Not all food is the same and neither are all feeders. If you are prepared to experiment with both, you may be surprised at the diversity of birds you can attract. But to attract the widest range of birds, you will need to provide food other than seeds (for example, fat, fruit, grated cheese), and use a bird table as well as feeders.*

Three other pieces of feeding advice. Don't just feed the birds in the winter: keep going during the summer too. Don't use those plastic net bags of peanuts that you sometimes see for sale as birds can easily get their feet stuck in the netting. Finally, having bought your bird feeder, make sure that it's topped up regularly – however bad the weather.

Don't forget about the food your garden provides for birds. Most gardeners are aware of the need to plant berrying shrubs (such as hawthorn, holly, pyracantha, cotoneaster, honeysuckle, ivy) for those birds, including blackbirds and thrushes, that like to eat fruit. Fewer gardeners actively cultivate plants to provide seeds for birds, but for those who do there is both good and bad news. The good news is that most plants produce seeds, and some birds will eat most of them – chaffinches are known to eat the seeds of over 100 different plants. The bad news is that most seed-eating birds seem actively to prefer the seeds of some pretty horrible weeds. Docks, knotgrass, fat hen, chickweed, groundsel and shepherd's purse are all high on their list of favourites. Beauty may be in the eye of the beholder, but the gardener has yet to be born who finds any of these plants attractive. In a large garden, you might be willing to set aside a corner for such undesirables, but for most of us some form of compromise is required.

To see what this compromise might look like, let's consider why so many seed-eating birds are in serious decline. The main problem is the loss of winter stubble, where birds could rely on finding a rich harvest of weed seeds and spilt grain to keep them going through the winter. Increased use of herbicides, more efficient grain harvesting and a national trend away from spring to winter cereals mean that winter stubble is now a rarity, and most farmland birds have declined along with their food supply. Gardeners can create their own miniature version of winter stubble by growing a patch of hardy annuals. For a really authentic look, mix in a few cereals – in my opinion, barley, rye and oats are quite attractive plants, although personally I'm not fond of wheat. For a more exotic look, there are other grain crops such as amaranth and buckwheat that also make quite attractive garden plants. Little management is required: allow all the plants to seed and die down naturally, and then pull up the remains of any larger plants and shake the ripe seed over the soil. Rake the soil lightly in early spring and pull up any perennial weeds that invade.

FOR YOUR OWN CORNFIELD:
- Cornflower *(Centaurea cyanus)*
- Corn marigold *(Chrysanthemum segetum)*
- Corncockle *(Agrostemma githago)*
- Marigold *(Calendula officinalis)*
- Poached egg plant *(Limnanthes douglasii)*
- Flax *(Linum grandiflorum)*
- Scarlet pimpernel *(Anagallis arvensis)*
- Field pansy *(Viola arvensis)*
- Love-in-a-mist *(Nigella damascena)*
- Honesty *(Lunaria annua)*
- Poppies *(Papaver commutatum, P. rhoeas, P. somniferum)*
- Quaking grass *(Briza maxima)*   continued ▶

- Foxtail barley *(Hordeum jubatum)*
- Barley *(Hordeum vulgare)*
- Oats *(Avena sativa)*
- Rye *(Secale cereale)*
- Grain amaranth *(Amaranthus cruentus, A. caudatus, A. hypochondriacus)*
- Buckwheat *(Fagopyrum esculentum)*

OTHER GOOD PLANTS FOR SEEDS FOR BIRDS:
- Evening primrose *(Oenothera spp.)*
- Forget-me-not *(Myosotis spp.)*
- Sunflower *(Helianthus annuus)*
- Teasel *(Dipsacus spp.)*
- Goldenrod *(Solidago spp.)*
- Knapweeds *(Centaurea scabiosa* and *C. nigra)*
- Michaelmas daisy *(Aster novi-belgii)*
- Milk thistle *(Silybum marianum)*
- Brook thistle *(Cirsium rivulare)*
- Tickseed *(Bidens ferulifolia)*
- Yarrow *(Achillea millefolium)*
- Rose *(any Rosa* as long as it has rosehips*)*
- Hawthorn *(Crataegus spp.)*
- Alder *(Alnus glutinosa)*
- Birch *(Betula spp.)*
- Larch *(Larix spp.)*
- Ash *(Fraxinus excelsior)*
- Pine *(Pinus spp.)*

Although these are all good plants, what you grow is less important than how you grow it. If one of your aims is to provide seed for birds, you must resist the temptation to deadhead.

Fruit and seeds, however, are only half the story. Many familiar garden birds eat insects, and even most of those that are quite happy with seeds and other scraps in the winter (such as blue tits, house sparrows and chaffinches) need insects during the breeding season. Young sparrows eat aphids, leafhoppers, beetles, flies, spiders, ants and caterpillars, and studies have shown that the more of these there are around, the more young sparrows survive. Try as you might, you are never going to provide the thousands of invertebrates needed by a growing family of sparrows or blue tits. Nor should you try – the main supply of these things must come from the garden itself. This supply depends largely on the volume of vegetation in your garden, and there is no substitute for covering every square inch with plants, and growing as many trees and large shrubs as you can.

Caterpillars are one of the main sources of food for birds, and moths are one of the main sources of caterpillars. Here are the top plants for moth caterpillars in Jennifer Owen's garden. Each is the food plant of at least five different moth species.

- Buddleja *(Buddleja davidii)*
- Shrubby potentilla *(Potentilla fruticosa)*
- Cherry plum *(Prunus cerasifera)*
- Roses *(Rosa spp.)*
- Spotted dead nettle *(Lamium maculatum)*
- Willows *(Salix spp.)*
- Red currant *(Ribes sanguineum)*
- Hawthorn *(Crataegus monogyna)*
- Marguerite *(Chrysanthemum frutescens)*

continued ▶

- Sweet marjoram *(Origanum majorana)*
- Michaelmas daisy *(Aster novi-belgii)*
- Perennial candytuft *(Iberis sempervirens)*
- Nettle *(Urtica dioica)*
- Parsley *(Petroselinum crispum)*

Other favourite foods for insect-eating birds are sawfly caterpillars and the larvae and adults of weevils and leaf-beetles. Hawthorns, birches and willows are good for all three.

You must also resist the urge to panic if some of your plants are attacked by a plague of caterpillars – remember, that's what you want *to happen.*

Spare a thought for flying insects. A few garden birds take insects on the wing (for example, swallow, swift, house martin and spotted flycatcher). For these birds, probably the best thing you can do is dig a pond, since many of the small craneflies, mosquitoes and midges that these birds eat have aquatic larvae. Many others breed in rotting vegetation, so a compost heap and generally plenty of organic matter will also help. And while we're on the subject of flying insects, please don't forget about food for bats. Those moth caterpillars that survive hungry blue tits will go on to provide food for bats. Because moths are the most conspicuous nocturnal insects, it's often assumed that they are the main food of bats, but this is true of only a few kinds. Most bats, including the pipistrelle, which is the species most likely to be seen in gardens, eat smaller insects such as various flies.

Before we leave the subject of food: don't forget to provide clean water, which birds need for drinking and bathing. Make sure that it's clean by regularly cleaning and rinsing the container. Keep it topped up in summer and replace with fresh if it freezes in winter.

NESTS

Award yourself one point for every bird that visits your garden, and an extra point for every one that finds something to eat, but score ten out of ten for every bird that successfully raises a brood of youngsters in your garden. These birds, which wouldn't have existed without your garden, are your real contribution to local biodiversity. However bad the weather or appalling the performance of your football team, the sight of a family of newly fledged blue tits is enough to lift the spirits of the grumpiest gardener. Nesting birds can also be surprisingly entertaining. I know some gardeners don't like them, but I never tire of seeing my local magpies strutting round the garden at nesting time, tugging at every twig they see and refusing to give up on one they really like, even if it's still firmly attached to the plant.

When most gardeners think of nesting birds, their thoughts turn naturally to nest boxes. Just as peanuts dominate the market for bird food, boxes for small hole-nesting birds dominate the market for nest boxes. The two main variations have either a small hole (25 millimetres), suitable for blue tits and coal tits, or a slightly larger hole (28–32 millimetres), suitable for great tits or sparrows. Larger boxes with larger holes might attract starlings, while jackdaws or tawny owls might use very large boxes. But most garden birds do not nest in holes, and not all those that do are keen on nest boxes. If you want nuthatches to nest in your garden, there's really no substitute for a real hole in a real tree.

Are birds nesting in your garden? If there are blue tits in your nest box it's hard not to notice, but don't assume that all nests are so easy to spot. Few other nests are quite as easy to find as those of blue tits, and the nests of some very common garden-nesting birds are downright difficult. Wren, dunnock and greenfinch nests are particularly well hidden. Jennifer Owen guessed from their behaviour that robins and dunnocks often nested in her garden, but she never succeeded in locating any nests of either.

A few other species, for example wrens, robins, wagtails and flycatchers, will use open-fronted nest boxes, but the majority of common garden birds do not use nest boxes at all. Some, such as the mistle thrush, goldcrest, magpie and goldfinch, usually nest in trees. But many frequently nest within about 2 metres of the ground, some of them (such as linnet and whitethroat) very close to the ground indeed. All like plenty of thick cover, and there is no substitute for a good dense tangle of brambles or ivy.

Before we leave the subject of nest boxes, a word about bats. Experience suggests that bats are pretty choosy, and that they use nest boxes only rarely and intermittently. If you see bats flying in your garden and are tempted to put up a bat box, be prepared to be patient, or disappointed, or both. There's also no guarantee that nest boxes for birds or bats will attract their intended occupants. Doubtless to the consternation of its owners, one bat box in a nudist colony in Hampshire became home to a colony of hornets.

Don't forget about birds with more unusual nests. If you see house martins or swallows flying over your garden, you might tempt them with special nests fixed under house eaves or against a beam in an open building. Alternatively you could make sure that there is a supply of the wet mud that both these birds need for nest-building in your garden.

> *If you want the birds that visit your garden to stay and nest, you must think beyond traditional nest boxes. At least ten species of birds nested in Jennifer Owen's garden, only one of them (blue tit) in a nest box. Plenty of shrubs and climbers, as impenetrable and thorny as possible, provide protection from predators and help to supply the insects and spiders that birds need during the breeding season.*

JOINED-UP GARDENING

Plenty of garden animals need quite different things from your garden at different times or at different stages of their life cycle. Pollinators such as hoverflies and moths, which need flowers for the adults and aphids or the right food plants for the larvae, are one obvious example. Birds are another. Resident birds may come to rely on your garden for food in winter and for nesting sites and food for their young in summer. Get one of these wrong, and your efforts to provide the others will also be much less successful than they could be.

Great tits are essentially deciduous woodland birds that have adapted well to gardens, and there is still a lot of coming and going between populations in gardens and in nearby woods. Woodland great tits have difficulty in finding enough to eat in winter, and winter survival is better in gardens where food is provided regularly. This is why most gardeners see great tits more often, and in larger numbers, in the winter. But for birds that are persuaded to spend the winter in gardens, there's a catch. Woodland provides a better food supply for raising chicks, and birds that stayed there all winter get first choice of territories and nesting sites in spring. So birds that hung around your bird feeder all winter may be tempted to stay and try to raise a family. One obvious way you can help is to provide a nest box, but this is only part of the answer. Typically, great tits raise fewer young in gardens than in woodland. This means that if you provided food all winter, and then tempted your resident great tits to stay by providing a next box, you have a moral obligation to try to provide the insects they need to raise their young. A bird feeder and a nest box alone, without the trees and shrubs to supply caterpillars and spiders, may simply mean that some of the young tits starve to death.

The primary concern of the bird-food and nest-box industry, not surprisingly, is to sell you their products. Nor is there anything wrong with this. Install feeders, bird tables and nest boxes in your garden and your local birds will undoubtedly benefit. But you should never forget that even if you buy everything in the catalogue, some of the needs of many birds (and all of the needs of some birds) can only be met by the garden itself.

THE BEST THING
SINCE SLICED BREAD

BY NOW, YOU MAY BE CONVINCED THAT GARDENS ARE THE best thing that ever happened to our native wildlife. Certainly a lot of intensive farmland would probably be improved, from a wildlife perspective, if it were replaced by houses and gardens. And yet there are great swathes of our native wildlife that are untouched by the activities of gardeners, and probably always will be. This may seem a depressing thing for a book on wildlife gardening to say – you want to know what your garden *can* do for wildlife, not what it *can't*. But the consistent theme of this book is that you will be a better wildlife gardener if you understand the wildlife in your garden – why it's there, what it's doing and what it needs. Part of this increased understanding is knowing the limits of wildlife gardening, and what you can realistically expect to find in your garden.

So let's begin at the bottom, with the basic green fabric of your garden.

PLANTS

We've had a lot to say about plants, but all of it about what plants can do for animals. What about your garden as a habitat for plants themselves? Are rare or interesting native plants to be found in gardens? Well, yes, up to a point. In the BUGS survey we found, among others, mezereon (*Daphne mezereum*), snakeshead fritillary (*Fritillaria meleagris*) and Jacob's ladder (*Polemonium caeruleum*), beautiful native plants that are uncommon or downright rare in the wild. Uncommon natives in Jennifer Owen's garden included clustered bellflower (*Campanula glomerata*), maiden pink (*Dianthus deltoides*) and sea holly (*Eryngium maritimum*). These plants may all seem very different, but they have one crucial thing in common: they were all planted. In other words, these plants are in gardens for exactly the same reason that there are tigers and elephants in a zoo – because someone put them there. I'm sure we all have different opinions of zoos, but I suspect we share a feeling that animals in zoos are somehow less important than those in the wild. If you were told the last wild tiger had died, but you could still see tigers at the zoo, you would be relieved that tigers weren't extinct, but I think you would also feel that a major conservation battle had been lost. I think you might feel much the same if mezereon were extinct in the wild but you could still see as many as you liked at your local garden centre. Is Cheddar pink quite the same plant in the rockery as it is on the limestone crags of Cheddar Gorge, and does Jacob's ladder look half as happy in the herbaceous border as it does carpeting the side of Lathkill Dale in the Peak District?

Nevertheless, plenty of plants do find their own way into gardens. What sort of plants are they? Owen found ninety-four 'volunteers' in her garden, overwhelmingly native weeds such as groundsel, ragwort, chickweed, fat hen and hairy bitter-cress. There was also a sprinkling of alien weeds, such as Canadian fleabane and American willowherb, plus some natives that are commonly planted but then tend to make themselves at home, such as hop, tutsan, creeping jenny and teasel. The BUGS survey supports this picture of the spontaneous flora of gardens: about a dozen native plants were more frequent in the BUGS gardens than any cultivated garden plant, nearly all of them either lawn grasses or weeds (for example, dandelion, willowherb, bramble, ragwort and creeping buttercup). Most of the other common native plants in gardens were also a pretty undesirable bunch, at least from a horticultural viewpoint: docks, plantains and hairy bitter-cress were all found in more than half our gardens. A few other natives, including foxglove, wood forget-me-not, dog violet and columbine, are actively planted or at least tolerated as garden plants. All readily self-seed.

Were there any native plants that found their own way into Owen's garden or the BUGS gardens that could be described as of any conservation significance? In a word, no. This is not to say that uncommon native plants don't turn up in gardens – they do, but not very often. In 1976 some gardeners on chalk in southern England, who had given up mowing their lawns during that year's severe drought, were astonished to discover that they had autumn lady's tresses (*Spiranthes spiralis*) in their lawns. Probably this orchid had been there for years, but frequent mowing had prevented flowering. Nevertheless, plants like this are the exceptions that prove the rule, which is that rare native plants find their own way into gardens only very rarely.

Why should this be so? There are at least two good reasons. The first is that if you look at the plants that are doing well in modern Britain, and compare them with those that are not, a clear pattern emerges. The winners are plants of fertile habitats (such as intensive pasture, arable fields and road verges), while the losers are plants of infertile habitats (heathland, chalk downs, ancient woodland). The massive use of artificial fertilizers and acid rain (a major source of nitrogen) are responsible for this pattern; plants that don't like fertile soils are drowning in a sea of nitrogen and phosphorus, while plants that like fertile soils have never had it so good. Gardens tend to be highly fertile – most gardeners, after all, spend a lot of time and money to make them that way – so they are a poor habitat for most rare plants. Rare plants can grow in fertile soils, of course; it's just that they are likely to be rapidly outcompeted by faster-growing weeds.

The second – and even more serious – problem is dispersal. As rare plants become more and more confined to isolated patches of older, 'unimproved' (that is, unfertilized) countryside, cut off from the mainstream of human activity, it gets harder and harder for them to escape. The power of normal methods of seed dispersal to overcome this obstacle is very limited. Orchids are much better at this than most plants, because of their extremely tiny, dust-like seeds. The sad truth is that, however impressed you are by the dispersal abilities of plants such as willowherbs and dandelion, this prowess owes more to being common in the first place than to the ability of any individual seed to travel long distances. Like many things, the barrier to dispersal only becomes obvious when it is suddenly removed. An example is the recent rapid spread of salt-marsh plants along the verges of main roads as a result of the use of de-icing salt. Salty road verges were available long before this sudden spread occurred, but the plants were unable to reach them. However, once just a few plants made it out on to the road network, they spread around the country almost instantly, aided by the most effective agent of dispersal in the modern landscape, the motor car.

So what does all this mean? It means that the average garden is an inhospitable place for most rare native plants. It also means that even if the right conditions exist in some gardens, a combination of isolation and poor dispersal prevents most rare plants from taking advantage of the opportunity. Nothing says rare plants can't make it into gardens, but none made it into Jennifer Owen's garden in Leicester or any of the sixty-one BUGS gardens in Sheffield.

This may be a good moment to reiterate a key point about wildlife gardening. However wildlife-friendly, it is still gardening. It's probably a subconscious feeling that successful wildlife gardening is about trying to duplicate or (worse still) replace the countryside that has led some people to place such emphasis on growing native plants.

Such duplication would be senseless even if it were possible, but in practice the attempt is futile. The semi-natural habitats that we value for their wildlife, such as lowland heath, old meadows, chalk grassland and ancient woodland, are the product of centuries of human management. Moreover, many of the common management practices that made them the way they are, such as flooding, burning and grazing, are simply impossible in the garden setting.

> *I would be wary of even trying to decide whether gardens are 'better' or 'worse' for wildlife than any semi-natural habitat, any more than whether heathlands are better than meadows. They're just different.*

Perhaps surprisingly, uncommon so-called 'lower' plants such as lichens and mosses (and their cousins the liverworts) are quite likely to turn up in gardens. Lichens, of course, aren't strictly plants at all, but a partnership between a fungus and an alga, which I suppose makes them half plant at least. Both derive a lot of their nutrients directly from rain rather than from soil and often grow on rocks or wood, while many are exceptionally well dispersed by spores. This relative insulation from fertile garden soils, combined with effective dispersal, means unusual members of both groups can show up almost anywhere, including gardens. Dramatic proof of this came in our discovery of what is almost certainly a new species of lichen (as yet unnamed) in the genus *Macentina*. Its preferred habitat appears to be tarmac, which perhaps accounts for it remaining undiscovered for so long. Another lichen found on a rock in a garden pond normally grows in mountain streams.

Many lichens were driven out of towns by air pollution, but cleaner air means that they are now back again, and a detailed search of the BUGS survey gardens revealed nearly 80 species, ranging from 2 to 30 per garden, with an average of about 15. Lichens are rather choosy about where they grow, and completely different species grow on acid rocks (for example granite or sandstone) and on limestone (or concrete). Large trees and shrubs provide good lichen habitat and so does weathered timber: half the lichens in our largest garden grew on a single old wooden bench, which was also home to two species that we didn't find anywhere else.

We didn't attempt to make a complete list of mosses and liverworts, but we still found sixty-eight species, some of them present in luxurious quantities in garden lawns. Although we didn't find anything nearly as rare as our new lichen, not all our species were common and a few were more typical of woodland or farmland. There seems to be no reason why a more detailed search shouldn't reveal some unusual mosses in gardens.

Grow rare native plants if you find them attractive, as many are, although you may find many of them hard to come by as plants or as seeds. Just don't expect any to turn up unannounced, however hard you try to provide the conditions they like. Even if they do, there's a serious practical problem, which is that most of these plants are slow-growing. If you leave every stray seedling until it's large enough to be sure what it is, your garden will soon be overwhelmed by fast-growing invaders that you probably already have, or don't want, or both.

The problem is very similar to the difficulty of establishing a flower-rich meadow on a fertile garden soil. The pretty wildflowers are easily swamped by fast-growing coarse grasses. If you want to grow most rare native flowers, they need protecting from this kind of thuggery.

ANIMALS

We're accustomed to some animals turning up in gardens more often than others, although the reasons for this are seldom obvious. For example, we take urban foxes for granted in Britain, but in the rest of Europe foxes are much less likely to be found in town gardens. Pine martens are shy and secretive animals, but their very close relatives, beech martens, are completely at home in towns and cities. So much so that they're often considered pests. Nevertheless, it's clear that most mammals and birds of sea shores, open water, mountains and moorlands are unlikely ever to find a home in gardens. But what about the invertebrates that make up 99 per cent of the biodiversity in your garden? When Jennifer Owen looked at the invertebrates in her garden as a proportion of the total number of species in the British Isles, she found quite a lot of variation between groups. She had almost half of all British harvestmen in her garden, nearly 40 per cent of the ladybirds and around a third of the butterflies, but only 20 per cent of the bees and 10 per cent of the grasshoppers. Which raises the obvious question: if you looked at enough gardens, would you eventually find all British butterflies and perhaps even all the grasshoppers too?

This is not a question anyone is likely to answer any time soon, and the BUGS survey wasn't exhaustive enough to allow us to say much about the total complement of animals in the BUGS gardens. But we can make some educated guesses, based on what we know about the garden habitat.

You are what you eat

The ecology and lifestyles of invertebrates are determined to a very large extent by what they eat. Consider, at one extreme, the burying or sexton beetles, large, handsome black and orange beetles that lay their eggs in carrion. This is very rich in protein and provides a complete, nutrient-rich food – the ultimate Atkins diet – for the beetle larvae. Accordingly, burying beetles have an extremely rapid life cycle – females from eggs laid today will be laying their own eggs just thirty days from now. At the opposite extreme, consider another family of beetles, the longhorns. Their larvae feed on dead wood, which is about the most useless diet you can imagine, so their larval development normally takes two to four years, but in dry, seasoned hardwoods may take much longer. Considering how far they need to travel in order to find fresh corpses of small mammals and birds, it's not surprising that burying beetles are extremely mobile, so they do visit gardens, although they are rarely seen. Jennifer Owen caught one or two individuals of three species in her garden. Not only are most longhorn beetles less mobile, but they don't often find the dead wood they need in gardens anyway. Although there are around seventy species of longhorns in Britain, Owen found only one species in her garden, and we found exactly the same species (but no others) in a handful of the BUGS gardens.

Carrion and dead wood are clearly dietary extremes, and it's hardly surprising that the insects that eat them don't have much in common, apart from being uncommon in gardens. You might expect insects that eat leaves to be more uniform, but not all leaves are the same. Try making a salad of oak leaves or pine needles and you will soon realize why both people and slugs prefer lettuce. Mostly we don't know much about how animals are affected by such subtle differences in diet, but in just a few cases we do. By far the best example is butterflies. All butterfly caterpillars eat plants and it turns out that, to a remarkable extent, the ecology of butterflies reflects the plants they eat.

At one extreme, butterflies that eat fast-growing, weedy plants like nettles, thistles and brassicas tend to be fast-growing and weedy themselves. They develop rapidly, emerge early, fly over a long period, often have several generations per year, and have wide geographical ranges and dense populations. Crucially, they are highly mobile and not of any conservation concern. The mobility of some of these butterflies is remarkable: painted ladies do not survive the winter in Britain and have to re-invade every year. From July to September, most of the painted ladies you see were born here, but most of those you see earlier in the year have flown all the way from North Africa. At the opposite extreme are butterflies whose caterpillars eat slow-growing plants of infertile habitats, which have exactly the opposite character-istics. Most importantly they tend to be immobile, and many are declining or endangered. The former group contains all the most familiar garden butterflies: small tortoiseshell, small and large whites, red admiral, painted lady, peacock, comma and orange tip. According to Butterfly Conservation, the first seven of these are consistently the most frequent butterflies in British gardens, while five of them were also much the most abundant in Jennifer Owen's garden. The latter group consists of butterflies that rarely if ever show up in gardens: silver-spotted skipper, adonis blue, pearl-bordered and dark-green fritillaries. These butterflies – and many like them – suffer from exactly the same problems as the rare native plants I mentioned earlier: isolation and low mobility. There's no mystery about their larval food plants: most of the fritillaries eat violets, the blues eat a variety of small members of the pea family and the skippers eat grasses (although the rare species are very choosy about exactly which grasses). You could grow all these plants in your garden quite easily, but your chances of attracting most of the butterflies that eat them are close to zero. Even a widespread and common butterfly in this group, such as the common blue, doesn't come anywhere near the top fifteen garden butterflies. In fifteen years Jennifer Owen found just two in her garden, while the whites and small tortoiseshells were numbered in thousands.

Butterflies are large, conspicuous and attractive, they fly during the day and there aren't many British species. To many gardeners, they're virtually honorary birds. Their populations have been monitored closely for at least 150 years, there are societies dedicated to their conservation, and we know much more about them than any other group of insects. To a unique degree, we know what makes them tick. Most other large insect groups, by comparison, are entomological black holes: for some rare British beetles, we still aren't quite sure where their larvae live or what they eat.

Nevertheless, there is every reason to assume that most large groups of animals occupy the same spectrum as plants and butterflies. At one extreme, fast-growing, adaptable, widespread, mobile species; at the other, slow-growing, specialist, uncommon, inflexible, slow-moving species. The former ('winners') are quite at home in the modern, fertile, disturbed landscape created by intensive farming and urbanization: nettles and cabbage white butterflies have never had it so good. The latter ('losers') have not liked a single thing that has happened to the British landscape since the Second World War, and are increasingly confined to fragments of an older, less intensively managed landscape. The former are very likely to find gardens congenial – in fact so congenial that some have become pests. The latter are very unlikely to find gardens at all, and aren't likely to find conditions to their liking even if, by some accident, they do end up there.

There may well have been 8,000 species of insect in Jennifer Owen's garden at one time or another (about a third of the British total), but the next 8,000 species may be quite a lot harder to please, while a significant minority are perhaps forever beyond the reach of all but the most remote and rural gardens. Still, 8,000 is a reasonable target to aim for, isn't it?

If you like violets and bird's-foot trefoil, by all means grow them, but don't expect the butterflies that eat them to make a beeline for your garden. The 'big six' (or seven) garden butterflies will probably show up whatever you do, but if you want to attract butterflies beyond the usual suspects, and especially if you want them to breed, try growing the larval food plants listed below.

Before you set out to grow any or all of these plants, one or two words of warning. Garlic mustard and hedge mustard are pretty *weedy, and the latter wouldn't win any prizes in a beauty contest: both are for 'wild' gardens only. Caterpillars of ten out of the sixteen butterflies listed eat grasses, and most are happy with lawn grasses, so letting some grass grow long has to be worth a try. However, if you want to be really helpful, try growing cock's-foot (*Dactylis glomerata*). This distinctive grass is unlikely to be in your lawn, but can be found in any road verge. Finally, all parts of both buckthorn and alder buckthorn are poisonous.*

The butterflies below are listed roughly in order of the chance of seeing them in gardens: you're more likely to see those near the top. Give yourself a pat on the back for every one of these butterflies you see in your garden. If you see any others, award yourself a gold star.

1. *Orange tip: garlic mustard (*Alliaria petiolata), *lady's smock or cuckoo flower (*Cardamine pratensis), *dame's violet (*Hesperis matronalis), *honesty.*
2. *Holly blue: holly and ivy (but bear in mind that the caterpillars feed on the fruits, so you must have a female holly and mature, flowering ivy).*
3. *Meadow brown, speckled wood, gatekeeper: long grass.*
4. *Brimstone: buckthorn (*Rhamnus cathartica) *or alder buckthorn (*Frangula alnus).*
5. *Green-veined white: garlic mustard, lady's smock, hedge mustard (*Sisymbrium officinale).*
6. *Small, large and Essex skippers, small heath, wall brown, ringlet, marbled white: long grass.*
7. *Small copper: sorrel (*Rumex acetosa).*
8. *Common blue: bird's-foot trefoil (*Lotus corniculatus).*

If you want to know more, visit Butterfly Conservation's excellent website (www.butterfly-conservation.org).

WHY YOU SHOULD CARE

THERE ARE PEOPLE WHO CAN'T SEE THE POINT OF WILDLIFE gardening. There are even people who can't see the point of gardening, although such philistines need not concern us here. Of course you don't belong to either group, but you probably know people who do. So, if you are forced to defend an interest in the welfare of butterflies or beetles, what do you say? In this short chapter, I'm going to give you two reasons to garden with wildlife in mind. Of course there are many others, but in my opinion these two are all you need. First, a big, global reason, and then a small, more personal one.

LOSING BIODIVERSITY

The planet is losing animals and plants at an unknown but alarming rate. Unknown because most of the species being lost are in the tropics, where species that don't even have names go extinct every day. Alarming because . . . well, there's a debate about just how much we should worry about biodiversity loss, but that's a different book. For the moment I'll assume we agree that conserving biodiversity is worthwhile, without enquiring too closely into motives.

How bad is the current biodiversity crisis? Very bad, for one simple reason. The human race is gobbling up unspoilt, pristine, 'wild' habitat at an ever-increasing rate. Already half the earth's land surface has been transformed by human activity, more than half the world's fresh water is used by man, and more atmospheric nitrogen is fixed by man (as fertilizer) than by all natural sources combined. And, like spoilt children, we simply mess up nearly as much of the planet as we actually use. On a global scale, the area currently under arable cultivation is about equal to the area of arable farmland abandoned since farming first began. Soil erosion is the major reason for abandoning land and most of this is now useless for crops and wildlife. In Australia, as native forest is felled, more rain percolates into the soil. Groundwater levels rise, bringing with them salt from deep in the soil profile. As salt accumulates at the soil surface, the land becomes useless either for the original vegetation or for crops. It is estimated that by 2050 17 million hectares of Australia will be threatened with destruction by this process, which cannot be reversed.

As land is used for living space or crops, or just ruined, the area available for the plants and animals that used to live there declines. Inevitably, species are lost. First to go are the rare, local and endemic species, which are lost more or less instantaneously. Later, and more slowly, commoner and more widespread species are lost. You don't have to look far to see this process at work – we have already driven about a quarter of the world's bird species to extinction. Ultimately, the relationship between loss of land and loss of species is brutally simple: if we lose 90 per cent of the original wild habitat, we lose 90 per cent of the species too. Thus even if 10 per cent of the earth's land surface is pristine nature reserve (in practice a tall order), we will still lose 90 per cent of all terrestrial species. Moreover, there are few causes for optimism. As far as biodiversity is concerned, most nature reserves are in the wrong place. Most species are in the tropics, but more than a third of the total protected area in the western hemisphere is in Alaska. At least double our present population now seems inevitable, and as the human population grows, we will need more living space, more food, more water, more wood. Finally, climate change means that many carefully preserved enclaves will no longer be suitable for the plants and animals they were designed to protect.

You could be forgiven for finding all this rather depressing. Yet there is a ray of hope. I have made the pessimistic assumption that land appropriated by mankind, for housing, crops or whatever, becomes quite unsuitable for its original complement of wild animals and plants. Of course this is far from true, and the challenge is to allow mankind to use more and more land, while at the same time making room for wildlife. This isn't as hard as it sounds – in fact many kinds of human activity are surprisingly compatible with wildlife. A golf course in Devon is almost the only site in Britain for the rare sand crocus. Many urban churchyards are de facto nature reserves. Although modern industrial-scale quarrying is completely unsustainable, many old abandoned quarries are now nature reserves. Military training is surprisingly good for wildlife. Not only does the army protect land from the twin evils of development and intensive agriculture but military training often actually benefits wildlife. On Salisbury Plain, a range of plants and animals of open habitats benefit from being occasionally run over by a Challenger tank. Military ranges in Britain are strongholds of everything from the chough to the stone curlew and the marsh fritillary butterfly to the green-winged orchid. Moreover the military have one big advantage over other landowners: they obey orders. At Fort Bragg in North Carolina, it's part of the 82nd Airborne Division's job to protect the Federally listed red-cockaded woodpecker, so they do. In order to protect the woodpecker, they have to protect its endangered longleaf pine habitat, so they do that too. Most private landowners have to be persuaded, cajoled or paid to protect wildlife, but the army just does what it's told.

Of course, the front line in the battle to make human land use compatible with wildlife isn't golf courses or military ranges. It's agriculture, which uses far more land than all other human activities combined. If the word agriculture conjures up a vision of a sterile prairie, don't forget that much of Britain's wildlife actually depends on low-intensity farming. Many of our most valuable habitats, from heathland to chalk grassland, are the product of centuries of grazing, burning and cultivation. Animals that really can't get on with a densely human-occupied landscape, such as the wolf, lynx and bear, went extinct centuries ago. There's therefore absolutely no reason, other than stupidity, carelessness or simple lack of will, why Britain should lose another single species. Even the government now recognizes that nature reserves are not enough, and that one vital ingredient of any long-term conservation strategy is more wildlife-friendly farming.

GARDENING AND CONSERVATION

In the midst of this belated concern over farming, gardens have tended to be forgotten. I've already pointed out that this is probably inevitable, since gardens are outside any form of statutory control and it's almost impossible to know what lives in them. Gardens are a vast resource, but if you don't know what's in them, and you couldn't do anything about it even if you did, you're likely to pretend they don't exist, and that's what the official attitude has been.

This official myopia has had some unfortunate consequences. Let me give you one example. You don't have to look into the conservation literature for very long, here or abroad, before you encounter the idea of 'wildlife corridors'. Such corridors are intended to help to counter the fragmentation that is all too common in the modern landscape. For example, hedgerows help to connect otherwise isolated woods and help plants, animals and birds to move between them. In rural areas, there's plenty of evidence that corridors work, but can the idea be extended to towns and cities? Here canals, rivers, railways and road verges are alleged to provide the arteries that connect otherwise isolated habitats and are essential if wildlife is to move around. The implications of corridors are plain: habitats on or near corridors should be richer in wildlife than those remote from corridors. A recent large-scale study in the West Midlands looked specifically for evidence of the effect of corridors. Plants and insects of numerous derelict brownfield sites were monitored and sites close to corridors were then compared to those distant from corridors. This large study found no evidence at all for any effect of wildlife corridors.

Like many unexpected results, this one is less surprising than it seems. The corridor concept assumes that corridors are wildlife-friendly routes through an otherwise hostile matrix. In intensively farmed countryside this is often the case, but in towns much of the matrix consists of gardens. Gardens are good wildlife habitat in themselves, and they join up to form a giant, linked network that clearly doesn't need much help from railways or canals. The emphasis on corridors not only distracts attention from the gardens that surround them but also makes wildlife gardeners who don't have a corridor at the bottom of the garden feel as if they are wasting their time.

You already know that having only a small garden, living in the middle of a city, not being particularly fond of native plants and having an unwillingness to grow nettles are no barrier to successful wildlife gardening. Wildlife corridors can now join the long list of things you don't need to worry about.

Over a fifteen-year period, up to a third of all known British invertebrates may have turned up in Jennifer Owen's garden at least once. In the previous chapter, I looked at this total as a glass two-thirds empty, choosing to emphasize the limitations of gardens as wildlife habitats. I'd now like to don my optimistic spectacles and take a more positive look at this number. Two questions spring to mind. How many species are there in Britain's 16 million gardens? We still have no real idea, but we do know that it's more than in Owen's garden. Consider that Owen found thirty-seven species of solitary bees in her garden. In the BUGS gardens, we found twenty-seven species, but five of them did not occur on Owen's list. In other words, even though BUGS barely scratched the surface of the diversity in a handful of gardens in a single English city, it added significantly to Owen's total. In the context of the whole of the UK, Sheffield and Leicester are rather similar, only an hour's drive apart. Who knows how many species might turn up in other towns and cities, in southern England, in Wales, Cornwall or Scotland?

Second, how many of the species in Owen's garden and in the BUGS gardens actually live and breed in gardens? For quite good reasons, Owen concluded that most of the species she caught were feeding and breeding in her garden. For all the more sedentary kinds (for example snails, slugs, worms, woodlice, centipedes, spiders, many beetles) this must be true. She thought that around a quarter of the flying insects were vagrants, chance visitors caught while just passing through. But passing from and to where? Owen's garden is far from rural, and it seems inevitable that virtually all the insects she caught were residents of gardens somewhere, even if not of her garden. The unavoidable conclusion is that, with few exceptions, the animals in Owen's garden and the BUGS gardens are real garden residents.

To sum up, up to a third of the entire UK invertebrate fauna occurred in Owen's garden; we can add significantly to that total with a much less intensive survey of a few Sheffield gardens; and most of these animals live and breed in gardens. And all this from looking at less than 0.001 per cent of British gardens.

What's more, although Owen's garden was managed very much with wildlife in mind, the BUGS gardens were selected for being very ordinary, and most were managed along very conventional lines. Although it's clear that gardens are already extremely rich in wildlife, and that they can provide a home for perhaps half of Britain's native invertebrates, we also know that much of the potential of gardens as wildlife habitats remains untapped. We know that compost heaps and ponds are great reservoirs of biodiversity, but that only a minority of gardens have either. We know that many species feed and breed in long grass, but that only a tiny number of gardens provide this habitat. We know that most gardens are a no-go area for the huge number of endangered insects that breed in dead wood. The inescapable conclusion is that the mid-term report on Britain's gardens is: 'Making excellent progress, but could do better'.

THE PERFECT WILDLIFE GARDEN?

How would you know if you had stumbled into the ideal wildlife garden? It would be bordered by tall hedges of mixed berrying shrubs, or perhaps a slightly dilapidated stone wall. It would be generally sunny and sheltered, but with a damp, shady corner. There would be a pond, with some long grass and a log pile or tree stump near by, a compost heap and at least one tree – probably two or three. There would be an earth bank or terrace, faced with recycled timber, bricks or stone. There would be lots of different plants and shrubs and little or no bare ground (apart from a patch of hardy annuals and maybe a vegetable plot), a rather weedy lawn, a tangle of climbers, and not too much paving or decking; the only chemical in the shed would be some garden lime. There would be one or two bird feeders and nest boxes, a dish of fresh drinking water, a bird table and perhaps a variety of home-made contraptions for nesting solitary bees and hibernating lacewings. There would probably be a faint but noticeable air of untidiness. Notice that none of this is incompatible with conventional, 'ordinary' gardening, provided you grow some periwinkle or a few nasturtiums over the log pile and rebrand the long grass as a wildflower meadow.

Such a garden would undoubtedly be a wildlife haven, and yet should we all aspire to gardens like this? Surprisingly, I think the answer is no, for two reasons. First, it's clear that this garden is trying to do everything, which may be a good idea if you have a large garden, but most of us don't. Moreover, as we try to fit more and more houses into our crowded island, gardens are only going to get smaller.

Second, much of the value of Britain's 16 million gardens is that every one is different. Consider the beetles found in litter and pitfall traps in the BUGS gardens. We found 167 different species, but very few were widely distributed. Only a handful of species occurred in more than twenty gardens, and half of them turned up only once. Spiders and several other groups showed a very similar pattern. In other words, a few species are common, but most are rare. You would be surprised how little we know about the basic biology of many of our native animals, and the smaller and rarer they are the less we know. Are all the less-common beetles and spiders distributed at random, with each garden's unique species drawn

from the pool like a lottery ball? Or was each species respond-
ing to some subtle aspect of the environment that made just
one of our gardens better than all the others, from the
perspective of that one species? We simply don't know, but
the possibility that the latter explanation is at least partly true
suggests that we shouldn't be too prescriptive about the ideal
wildlife garden. Maybe somewhere there's an insect that
thrives on decking, stainless steel and crushed blue glass,
nourished by just the odd sedge or tree fern.

> *Don't despair if you have neither the space nor the inclination to*
> *provide all the ingredients of the ideal wildlife garden. If you*
> *haven't the time to dig a pond, don't worry: there's probably*
> *one in the garden next door. If you haven't room for a meadow,*
> *that doesn't matter as long as there's one in the neighbourhood*
> *somewhere. In fact probably the best advice is to try to contribute*
> *something that none of your neighbours is providing. The only*
> *golden rule is not to use pesticides. Beyond that, it's up to you*
> *whether you make a great or a little effort to attract wildlife,*
> *or even none at all. Either way, plenty of wildlife will still enjoy*
> *your garden.*

OVER THE GARDEN WALL

If you've got this far, I think it's fair to assume that we share a
vision of a future in which Britain's gardens have become
even better for wildlife than they are already. But if you really
care about our native wildlife, make sure that your concern

doesn't end at your own garden fence. Consider the impact of your gardening on wildlife in the wider world.

Don't use peat or peat-based compost, even indirectly. Peat is a non-renewable resource and peat bogs support a diverse community of animals and plants, many of which live nowhere else. Patronize local nurseries that are committed to reducing peat usage and that propagate and grow their own plants, rather than trucking them in from the other end of the country or from The Netherlands. If you use a garden centre that uses peat-based compost, ask them why. Preferably use local timber and use only tropical hardwoods from certified sustainable sources. Use recycled timber, bricks and stone whenever possible. Don't buy wild-collected plants or bulbs.

Compost or shred all the garden waste that can be composted or shredded; larger woody material can go on the log pile. Garden waste contributes to air pollution if it's incinerated, or to the destruction of natural habitats if it goes to landfill. Think twice before buying thoroughly unnecessary (and noisy, energy-wasting or polluting) gadgets like leaf blowers, patio heaters and powered lawn scarifiers. Don't waste money on an expensive subscription to your local gym and buy yourself a manual lawnmower instead. Don't waste water – get yourself a water butt and use it. Don't automatically reach for the lawn sprinkler every time it doesn't rain for a week – however brown your lawn looks, it will recover when it rains in the autumn. Use fertilizers sparingly and use animal manures or home-made compost whenever possible. Manufacturing artificial fertilizers uses a lot of energy and they are much more likely than natural manures to contaminate watercourses and groundwater. Buy local charcoal for your barbecue. Not only is much imported charcoal from illegally logged forest: buying locally provides financial incentives for the better care of British woodlands.

RECONNECTING WITH NATURE

I promised you two good reasons for gardening for wildlife, and here is the second one. Half the world's population lives in towns and cities. In Britain the proportion is more like 90 per cent. For most of us, food comes shrink-wrapped from the supermarket and the closest we get to wildlife is a David Attenborough documentary. Yet for nearly all our evolution, we were 'wildlife' ourselves, and human beings retain an innate affection for living things. The great American naturalist Edward Wilson has called this affection 'biophilia'. We are all born with an instinctive biophilia, which can be either stifled or nurtured by our early experiences. In the absence of any real intimacy with the living world, it's easy to become indifferent to the fate of the other creatures with which we share our planet. How can someone who has never seen a sparrowhawk be expected to care about the fate of the condor?

There was a time when many young people made early contact with nature via what Richard Mabey christened the 'unofficial countryside' – abandoned industrial land and urban wasteland. Free from paths, fences and 'keep off the grass' signs, such habitats provided the intimate contact with nature largely withheld by parks, zoos and nature reserves.

Denied this experience by real or imagined parental fears and the seductive lure of the internet and computer games, many children now grow up without forging any close personal bond with their local flora and fauna. As Rachel Carson pointed out, exploration of the natural world is far more than just a pleasant way to pass a few childhood hours. If one grows up lacking even a rudimentary knowledge of the way of life of non-human neighbours, the natural world can lose the power to arouse fascination, wonder and awe. It's then all too easy to believe the propaganda of the chemical industry and allow a fixation with the few genuine pest species to develop into a hatred of creepy-crawlies in general and insects in particular. But this is absurd. Not everything coloured black and yellow will sting you, not every small fly is a mosquito, and not every caterpillar is out to destroy your cabbages or roses. Around a third of all *known* animal species are insects, and the true proportion is certainly higher. If you hate insects, you hate life.

What can be done to reconnect people with the natural world? One thing is plain: the nature conservation establishment thinks the answer doesn't have a lot to do with gardens. Natural England, the government's nature conservation agency in England, recently discussed a draft strategy for urban biodiversity. It's very much concerned with reconnecting people with nature, but only one of its eighteen targets mentions gardens, and this is only to propose that there should be more research on the value of gardens for biodiversity. Instead the emphasis is on accessible green space and statutory local nature reserves. But the accessible green space is likely to be the local park, and parks – with a few honourable exceptions – are far from wildlife oases. Don't misunderstand me. Everyone *should* have access to open space to walk the dog, play football or fly a kite, but the typical urban park is not the place to make intimate acquaintance with wildlife.

Nor are nature reserves necessarily the answer. When did you last visit your nearest local nature reserve? How many of us even know where it is? Nature reserves bring us back to a central debate about nature conservation, which is why conserve? Ever since the establishment of the Nature Conservancy in 1949, this has been essentially a scientific question, with scientific answers. Nature reserves were seen very much as outdoor laboratories for the study of nature, and even the urban nature movement of the 1980s did nothing to break this mould. Nature reserves are places you're allowed to visit, preferably in the company of an expert. The idea that nature reserves are there for people to enjoy, and that we all benefit spiritually, emotionally and physically when nature is accessible, has never really caught on.

In fact anyone with a garden has their own nature reserve, literally on the doorstep. What's more, there are no restrictions on access to this reserve, day or night, rain or shine. You can develop a close personal relationship with your wildlife and observe its changing with the seasons. Begin with the obvious wildlife – the birds, butterflies and bumblebees. Sit or stroll quietly and soon you will begin to see the less obvious animals too: the hoverflies, the solitary bees, the parasitic wasps stalking their prey, the ladybird larvae hoovering up aphids, a spider spinning a web. Of course you'll see animals eating your plants, but you can learn to understand and tolerate them. If you really can't bear to see your plants eaten by slugs, grow plants that slugs don't like, which will cause both you and the slugs less trouble in the end. I've quoted Jennifer Owen on this before, but it's worth repeating: 'There are no pests, because everything in my garden is a source of interest and enjoyment.'

ACKNOWLEDGEMENTS

Some brief thanks are necessary. First Kevin Gaston, Phil Warren, Richard Smith and Alison Loram, without whom much of the first half of this book could not have been written. If any of my facts turn out to be correct, they can take much of the credit. If any of my opinions turn out to be wrong, that's my fault. Second, everyone else involved in the Sheffield BUGS project, not least the scores of garden owners who kindly allowed us the use of their gardens. Third, the Natural Environment Research Council for generously funding the BUGS project. Fourth, Jennifer Owen for continuing inspiration. Fifth, everyone at Transworld for their customary friendliness, enthusiasm and efficiency. Finally thanks to Pat, Lewis and Rowan for their continuing support and encouragement during my transition from 'unknown academic' to 'unknown academic with a small sideline in gardening books'.

BUGS PUBLICATIONS

Popular publications

Thompson, K. (2004), 'Bugs in the Borders', *The Garden*, 129, 346–9.

Gaston, K. J., Smith, R. M., Thompson, K. and Warren, P. H. (2004), 'Gardens and Wildlife – the BUGS Project', *British Wildlife*, 16, 1–10.

Thompson, K. (2006), 'Cashing in on the Birds and the Bees', *The Garden*, 131, 356–357.

Scientific publications

Thompson, K., Austin, K. C., Smith, R. M., Warren, P. H., Angold, P. G. and Gaston, K. J. (2003), 'Urban Domestic Gardens I: Putting Small-scale Plant Diversity in Context', *Journal of Vegetation Science* 14, 71–8.

Gaston, K. J., Smith, R. M., Thompson, K. and Warren, P. H. (2005), 'Urban Domestic Gardens II: Experimental Tests of Methods for Increasing Biodiversity', *Biodiversity and Conservation* 14, 395–413.

Thompson, K., Hodgson, J. G., Smith, R. M., Warren, P. H. and Gaston, K. J. (2004), 'Urban Domestic Gardens III:

Composition and Diversity of Lawn Floras', *Journal of Vegetation Science* 15, 373–8.

Gaston, K. J., Warren, P. H., Thompson, K. and Smith, R. M. (2005), 'Urban Domestic Gardens IV: The Extent of the Resource and its Associated Features', *Biodiversity and Conservation* 14, 3327–3349.

Smith, R. M., Gaston, K. J., Warren, P. H., and Thompson, K. (2005), 'Urban Domestic Gardens V: Relationships Between Housing, Landscape and Habitat Composition', *Landscape Ecology* 20, 235–53.

Smith, R. M., Warren, P. H., Thompson, K. and Gaston, K. J. (2006), 'Urban Domestic Gardens VI: Environmental Correlates of Invertebrate Species Richness', *Biodiversity and Conservation* 15, 2415–2438.

Thompson, K., Colsell, S., Carpenter, J., Smith, R. M., Warren, P. H. and Gaston, K. J. (2005), 'Urban Domestic Gardens VII: A Preliminary Survey of Soil Seed Banks', *Seed Science Research* 15, 133–41.

Smith, R. M., Gaston, K. J., Warren, P. H. and Thompson, K. (2006), 'Urban Domestic Gardens VIII: Environmental Correlates of Invertebrate Abundance', *Biodiversity and Conservation* 15, 2515–2545.

Smith, R. M., Thompson, K., Hodgson, J. G., Warren, P. H. and Gaston, K. J. (2006) 'Urban Domestic Gardens IX: Composition and Richness of the Vascular Plant Flora and Implications for Native Biodiversity', *Biological Conservation* 129, 312–322.

INDEX

If you enjoyed *No Nettles Required: the reassuring truth about wildlife gardening*, then you will also enjoy:

*An Ear to the Ground:
garden science for ordinary mortals*
by Ken Thompson

'Good popular-science books for gardeners do not come along every day. Those that are easy to read, witty and do not insult the intelligence of readers are rarer still. So I can't recommend too highly *An Ear to the Ground* by Ken Thompson, which gives up-to-the-minute answers to all those questions that nag at you when you are bottom-up in a flower border or floating off to sleep at night.'
Ursula Buchan, *Daily Telegraph*

'Wise and refreshing.'
The Garden

'This joyful little book will help debunk some gardening myths, and reveal your garden as it really is.'
Gardening Which?

'Peace of mind for anxious gardeners.' *Daily Mail*

'An excellent read.' *The Times*

ISBN 1903 919193
Hardback non-fiction
£10.00

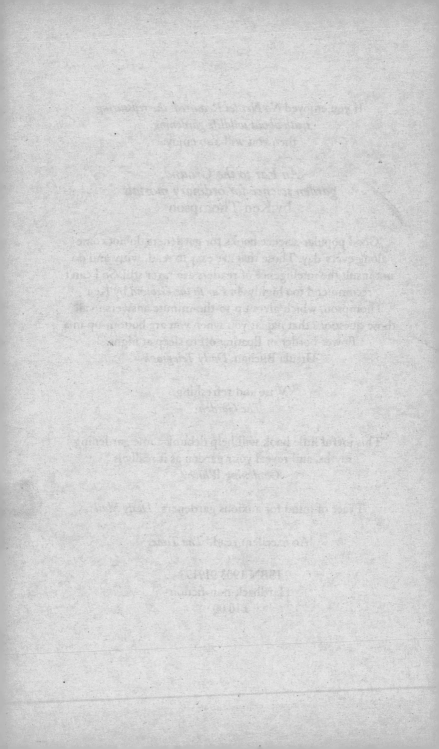